Before the Beginning of Time

Before the Beginning of Time

Jacob Meyerowitz

rRp publishers · Easton

Grateful acknowledgment is made for permission to reprint material from the following:
Journal of Orgonomy, Princeton, NJ: "Basic Orgonometry," 19(1), 1985; "Index of Orgonometric Symbols," 19(2), 1985; "Function as a CFP," 19(2), 1985; "The Formulation of *Right* and *Partly Right*," 20(1), 1986; "Language and the Development of Functions," 20(2), 1986; "Hegel's Dialectic Concept: An Orgonometric Review," 21(1), 1987; "The Function that Defines the Goal," 22(1), 1988; "How to Integrate an Unknown Function," 23(1), 1989; "The Orders of Function," 23(2), 1989; "The Primordial Universe," 24(2), 1990; "The Function of the Orgasm: An Orgonometric Review," 25(2), 1991; "The Source of Time and Length," 26(1), 1992.
Applied Orgonometry, Easton, PA: "The Real Meaning of $E = mc^2$," No. 1, March 1986; "Art as a Process," No. 2, November 1986.

 Published in the United States by RRP Publishers, 5 North Bank Street, Easton, Pennsylvania 18042.

Library of Congress Catalog Card Number: 93-86657
ISBN: 0-9607034-2-X

First Edition: January 1994
Paperback Edition: September 2000
Printed in the United States of America.

CONTENTS

Dedicated to expanding the
knowledge uncovered by
Wilhelm Reich,
the inventor of orgonometry.

PREFACE AND ACKNOWLEDGMENTS

Whoever has himself worked in unexplored territories understands that it is not a predetermined goal, but the way of searching which itself is reflected in the end product.

WILHELM REICH (1)

Orgonometry expresses a new way of thinking, derived from observing how things actually work, that is, from the logic of functions. The way things work and the way we *think* they work do not follow each other precisely. At best, there has always been a small, previously undefined discrepancy. Many scientists believe that "pure" logic will eventually resolve this little discrepancy. They, as well as others, have assumed, given the facts, the irrefragable facts, that reason and logic will eventually uncover how everything works and how everything is integrated.

Unfortunately, this optimistic view of reason and "pure" logic is an error of thought. The discrepancy cannot be resolved by improved measurements or better formulas, because it is the symptom of a built-in, systemic misalignment. The conclusions in this book will repeatedly show that the old way of thinking and the way nature functions are always noticeably misaligned. It follows that this misalignment must be entirely due to the logic of thought. We cannot blame the logic of functions for not conforming to our thoughts. Functional thinking, that is, the new way of thinking, dissolves this problem, because it is based upon the logic of functions and not upon the logic of thought.

This book will always be the first of its kind — the first to describe original research guided by the technique of orgonometry and defined with orgonometric formulations. Its contents reflect the many directions of my interest and the broad range of my practical experience. The variety of the subjects intentionally disregards the artificial boundaries of established disciplines, because the new way of thinking and the technique of orgonometry intrinsically reunite and integrate knowledge. As a whole, this collection of seemingly unrelated subjects will serve to illustrate the universal application of the technique of orgonometry and the comprehensiveness of the new way of thinking.

The reader should not be discouraged by the appearance of formulations within the text. This book is *not* concerned with mathematics; nor is it directed toward any specific "compartment" of knowledge. Although scientific subjects predominate, they are interspersed with non-scientific subjects. All of them are relevant examples of functional thinking. The chapter on Art uncovers the underlying function of original art; the chapter on Language demonstrates the practical limit of all languages; the one on Hegel's Dialectic illustrates the limit of "pure" thought. These applications could not be described with classical mathematics.

The equations that appear in this book are formal descriptions of qualitative relationships which, after a little practice, can be read like diagrammed sentences. Everything the reader requires to follow the logic and to understand the meaning of my conclusions is introduced and developed in the first two chapters.

After several of my papers were completed, I began to think about them as possible chapters of a future book. Each paper described the results of an investigation of a specific subject, and each presentation was treated as an autonomous whole. Practically, they were all linked by a common approach: Functional thinking and the use of the technique of orgonometry. I was certain the common approach itself would unify the various subjects represented, but found they were even more integrated than I had anticipated. In addition to all the cross-references and other technical unities, the chronologic sequence conveys a sense of steady progress — the development of my comprehension — which I could not have anticipated. This unexpected *whole* expression has energized me to prepare this book without delay, because new formulations have accumulated, and there is no end to what can be uncovered with this technique.

Working with orgonometry has been a lonely occupation. This did not bother me at first, because the work itself was so fascinating and all-absorbing. However, my feeling of being isolated in my work increased as I accumulated more and more functional conclusions about an array of different subjects. Many of these conclusions contradict accepted facts or theories. I often wished I could talk to someone about these strange, new conclusions, without having to explain the fundamentals of orgonometry each time. More and more, I came to rely on the writings of Wilhelm Reich to support my conclusions. No one has ever been as consistently functional in his thinking as Reich was.

There is a difficult aspect to this work, known only to those who work in "unexplored territories." Toward the end of an investigation that leads to the discovery of an entirely new functional relationship, especially one that contradicts accepted knowledge, the investigator is thrust into a disquieting state. He finds himself alone in an uncharted landscape and must be guided entirely by his own perceptions. In the past, the only principle we had to direct us was the concept of *one* God. Now we have the logic of functions and the technique of orgonometry to guide us. This is a great advance for the exploration of the unknown. Nevertheless, what the investigator perceives, as he steers himself through this uncharted landscape, may still be limited by what he, as an individual organism, can tolerate or express. Personal anxieties or fears can divert even the best-equipped explorers.

Acknowledgments

Although my work is mainly accomplished in solitude, a number of persons have assisted me in the preparation of my manuscripts. Others have helped to promote my work, which in my view, means promoting orgonometry. My wife, Patricia, has consistently supported and helped me. Because of her deep interest, she has assumed many roles: My first audience, my first reader, and my first editor. Her questions and criticisms have always been significant, and my descriptive writing is constantly being improved because of her help. She has also acted as the agent for this book. I cannot thank her enough.

I owe a special debt to Linda Ketron, who volunteered her expert help as an editor and later undertook the equally arduous task of preparing the camera-ready pages for the publisher. Her practical skills and attention to detail undoubtedly led to further improvements in my text, and her role in the final stages of preparing this book for publication made the whole task more pleasurable.

As author, the text and its meaning absorbed all my attention. I would have gladly left the task of designing this book to others. However, the technical process of reproducing the visual qualities of all the orgonometric formulations inevitably propelled me into organizing the design of the whole book. In this capacity, I was ably assisted by Gary M. Smith, an artist and computer-graphics expert, who prepared all the symbols, illustrations, and most of the equations that guided my text. Thank you, Gary.

Thanks also to the editorial staff of the *Journal of Orgonomy* for indirectly helping with this book. They read and edited my papers before they were originally published. Among them, the former Managing Editor, Linda Ketron (mentioned previously), was exceptionally helpful in overseeing the correct presentation of the equations. This task alone has been frustratingly difficult at times. Her successor, Kathleen Erickson, has been equally helpful and meticulous.

Paul Mathews and John Bell were the first to invite me to speak about my work with orgonometry, and they encouraged and supported me thereafter. The late Paul Mathews was impressed with my new conclusions and incorporated them into his own lectures. The late Elsworth Baker, founding editor of the *Journal of Orgonomy*, became a strong supporter in 1984. After reading the manuscript for "Basic Orgonometry," he concluded that the technique was very important and my work must be published. He encouraged me to write about it, and what followed was the steady stream of papers that constitute the greater part of this book. The current editor, Richard Blasband, has also become a strong supporter of orgonometry. Under his direction, my tenth paper was presented as the lead article in the *Journal of Orgonomy*.

Others who have helped at various times are: Howard Chavis, Robert Harman, Charles Konia, Barbara Koopman, Robert Pasotti, Richard Schwartzman, Myron Sharaf, and Argyro Collins, my voluntary research assistant. I thank them all. Last but by no means least, I thank my brother Dave for his support and loving generosity.

Easton, PA   February 1993

1

ENTER THE LOGIC OF FUNCTIONS

Sections

1. Diagramming the Development of the Amoeba
2. The Meaning of "Function"
3. Preparatory Considerations
4. The Pattern of the Development of Functions
5. The Pattern of Paired Functions
6. The Common Functioning Principle
7. A Summary of the Given Propositions
8. A Visual Demonstration of Development
9. Looking Toward the Right — the Direction of Development
10. Looking Toward the Left — an Artificial View
11. The Source of Everything
12. Fragments of Functional Thinking in Our Past

If you learn of a new basic function in nature, be ready to revise your well set ideas.

WILHELM REICH (1)

Initially, we must assume the reader has no prior knowledge of the principles and the method of functional thinking. Nevertheless, most of us will have had some previous experience with this kind of thinking. Many practical tasks cannot be accomplished without an input of functional thinking. As we proceed with the content of this book, some of us may well recall such past experiences.

In my youth, I was confronted with two descriptions of bird flight:

- Birds developed wings in order to fly;
- Birds fly because they have wings.

I recognized both of these statements were false, and furthermore, they represented a way of thinking that does not adequately describe the development of birds, wings, or flight. I mention this because those observations were my first conscious experience of the existence of a different kind of thinking — the kind we now call "functional thinking."

Functional thinking differs from the familiar way of thinking. Its *logic* is derived from the operations of function rather than from the operations of thought. For example, the "direct" relationship between *cause* and *effect* is an idea that comes from the logic of thought, not from the logic of functions. According to the logic of functions, *cause* does not determine an *effect*. The association of *cause* and *effect* is determined by another, unstated, single factor more profound than either *cause*, *effect*, or both of them together. This means the expressions of *cause* and *effect* must be simultaneously *identical* and *different*. In practice, we often encounter an "effect" that becomes the *cause* of something else, or a "cause" of something later seen to be the *effect* of something else. In such cases, the accepted meanings of *effect* and *cause* appear to be not only identical but technically interchangeable.

Experiencing the logic of a development of functions seems the best way to enter the realm of functional logic. I believe we can induce this experience with visual aids in a simple but stimulating experiment of thought designed expressly for the task. However, we will need some technical briefing before we can proceed with the experiment of thought. This too will be presented by way of an experience. Hopefully, the material in this chapter will prepare the reader for the kind of thinking that originally generated the technique described in the next chapter.

1. Diagramming the Development of the Amoeba

To orient ourselves, we begin with a seemingly ordinary demonstration of how we can diagram a commonplace living process. The propagation of living cells, such as the amoeba, can be represented with the following diagram:

AMOEBA 1 → **AMOEBA 2** → **AMOEBA 4**, **AMOEBA 5**
AMOEBA 1 → **AMOEBA 3** → **AMOEBA 6**, **AMOEBA 7**
→ **etc.** 1.01

Some time after it has matured, **AMOEBA 1** splits apart to become **AMOEBA 2** and **AMOEBA 3**. **AMOEBA 2** and **AMOEBA 3** gradually mature and expand, and each one likewise splits in two becoming **AMOEBAE 4** and **5**, and **AMOEBAE 6** and **7**, respectively. If all goes well, each **AMOEBA** of the third generation will also divide to become two **AMOEBAE**, and this process (mitosis) goes on and on.

By substituting a common letter-label for all the constituent cells in the above diagram, we generalize the whole arrangement. In this more abstract form, our diagram can be used to represent the basic process of cell propagation for all living species:

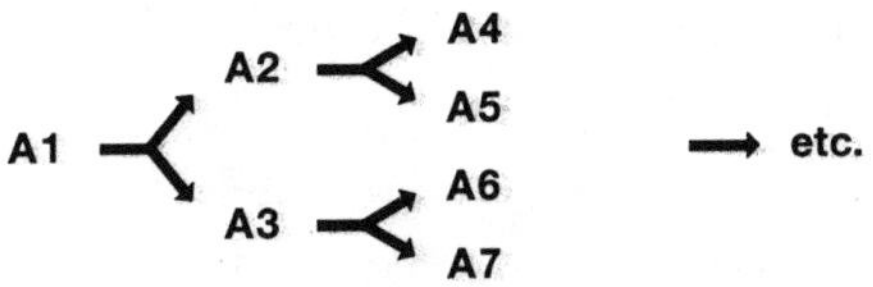

A4
A2
A5
A1 → **etc.**
A6
A3
A7 1.02

The diagram illustrates the general form of the *development* or propagation of living cells, including the amoeba. **A1** represents the parent cell; **A2** and **A3** the second generation of cells; **A4**, **A5**, **A6**, and **A7** the third generation of cells.

In the above context, the letter-label **A** represents the functioning principle of a species and is therefore common to all the offspring. The number-label differentiates the individual amoebae or constituent functions. The arrow on the right indicates this kind of development continues beyond the right-hand end of the diagram. There is no known limit to the extension of such a development.

2. The Meaning of "Function"

There is nothing mysterious about the way we use the term "function," but admittedly, this word is associated with a seemingly "mysterious" experience. Our grasp of its meaning fluctuates in keeping with fluctuations in our biophysical responses and perceptions. Consequently, we need to review its meaning from time to time. We use the noun "function" to mean: The *role* of anything; or the *work, working, activity* of anything. This meaning is not specialized but conforms with the generally accepted definition of the term.

3. Preparatory Considerations

The first requirement of the new way of thinking is to observe how things function; the second, to describe everything according to the way it functions. To fulfill these requirements, we need to consider how we perceive and differentiate functions, and how we can best describe functions.

These considerations can lead us to observe the intimate relationship that exists between our perceptions of quality and our capacity for distinguishing different functions. We would then recognize that functions are immediately perceived as qualities, not as quantities. This does not mean functions cannot be quantified, but that perception is essentially a qualitative experience.

The quality of individual functions can either be named or represented with abstract letter-labels. It is better to begin with the names of functions rather than to abstract them. Name-labels are significantly more flexible than abstract labels. Abstract labels require more detailed, initial knowledge, because they transmit more complete information than name-labels.

Since quality distinguishes functions, quality is of paramount importance for the new way of thinking. The old way of thinking is carefully structured to avoid "subjective" responses, whereas the new way requires us to take a chance with the "subjective," for a very important reason. By eliminating the "subjective," the old way of thinking sharply limits what it can encompass. By including the "subjective," the new way becomes integrated with nature, thereby making it more penetrating and more profound. It is reasonable to assume the operations of our sense organs and the process of perception, which originate in nature, must operate like other basic functions to be found in the universe.

Errors of perception will occur, but the new way of thinking is stimulated by encounters with such errors. Discovering an error is a sign of progress in comprehension, which in turn can lead to further discoveries. However, the entire content of perception should not be excluded simply because it can err.

4. The Pattern of the Development of Functions

In practice, verbal descriptions of functional processes can often complicate rather than clarify the operations of functions. The reason in brief: The logic of words in the organization of a sentence conflicts with the logic of functions in the expression of a development. Since what is to be illustrated is essentially quite simple, I turn to the use of diagrams as the most direct way to convey this kind of simplicity.

The common operation, or pattern, that emerges from the previous diagrams can be represented by the form of the following symbols:

 1.03

Since this graphic symbol expresses the operation of a function, it is meaningless without the inclusion of the determining function and the pair of functions it generates. All three functions can be represented in the abstract as follows:

A2
A1
A3 1.04

This diagrammatic arrangement represents the most basic pattern of the relationship of functions. It describes the development of one function into two functions, and it demonstrates how two individual functions, **A2** and **A3**, are related by their common "parent" function, **A1**.

The basic pattern above can be used to describe the development of every kind of function. However, different constituent functions may require a different set of letter-labels carefully chosen to reflect whether or not the constituents are of the same kind. This technical matter will be fully discussed in the next chapter. Here, to complete the picture, we need only show one example of a development with a different mix of constituents:

B1
A
B2 1.05

In this development diagram, the common function **A** generates two variations of a different kind. The pair of variations **B1** and **B2** are of the same kind. Though the pattern of the development is the same as before, the whole process is otherwise different.

5. The Pattern of Paired Functions

Historically, the paired relation of functions was the first functional pattern to be recognized. The above one-becomes-two pattern — the pattern of development — was only grasped after a body of evidence confirmed that every function is paired with a different yet simultaneously identical function. This strange but correct description of the relation of paired functions remained a puzzle, until a momentous act of thought discovered it to be logical in the broader context of the development of functions.

Given the present accumulation of functional evidence, we can readily show that the original observation about the natural pairing of functions is not merely a "floating thought" or a "philosophic concept," but a concrete description of the operations of objective functions. For example, using the amoeba again as a reference, the two offspring of one amoeba, **A2** and **A3**, are obviously related and obviously a pair (i.e. the equivalent of twin sisters). This paired variation can be represented as follows:

A2 — A3 1.06

Amoeba **A2** and amoeba **A3** are related as a pair of simple variations. Both creatures are alike and each is also similar to its parent. But what really ties them together functionally is their common parent, amoeba **A1**, which is not shown in this monolinear arrangement.

We could list numerous other examples to prove that functions are generally related as pairs. Electricity is expressed as *positive* and *negative*, magnetism as *north* and *south* polarities, *dynamos* can be transformed into *motors*, *expansion* alternates with *contraction*, *charge* is followed by *discharge*, *male* is attracted to *female*, *pleasure* is opposed by *anxiety*, and *plants* are associated with *animals*. The relationship of paired functions ranges from simple variations through antithetical pairs to functions of different kinds.

Most functional pairs are not related as simple variations. The majority are opposites of varying degrees. For example, magnetic *north* and *south* are related as mutually attractive opposites, *male* and *female* are also mutually attractive opposites, whereas *expansion* and *contraction* are alternating opposites.

6. The Common Functioning Principle

If we had been the first to discover that *every function is directly associated with another function* and, in most cases, *the two qualities are antithetical*, we would then be confronted with a crucial question: What real or objective factor determines this association? Or phrased another way: What concrete quality identifies the two functions as a functional pair? To answer this question under those conditions required a creative leap of thought few of us can hope to match.

Because the above question has been resolved, we can presently find the answer in the basic diagram for the development of functions presented previously (1.04). In this diagram, function **A1** (on the left) determines the association of the paired functions **A2** and **A3** (on the right). The function on the left is the source of both constituents of the pair of functions on the right. The function on the left is almost always deeper, broader, and more profound than the functions it determines (on the right). In this way, development generates a natural order of functions.

We understand now that paired functions, no matter how dissimilar they may be in some ways, are simultaneously identical with respect to the specific function that determines them both. In the abstract, we refer to the role of this deeper function as the *common functioning principle*, or CFP.

7. A Summary of the Given Propositions

To exercise the logic of functions and follow it to its furthermost reaches, we must first be convinced about the basic observations from which this logic is derived. Since this is not our present aim, let us simply *assume*, for the duration of the next demonstration, that the following four propositions about the basic operations of functions are indeed correct:

- Every function is directly associated with another function.
- The association of paired functions is determined by a third function.
- Functions develop according to the simple bifurcating pattern described previously (also loosely described as one-becomes-two, or one-determines-two).
- The function on the left of a development (i.e. the CFP) is technically deeper or more profound than any of the functions to its right (i.e. the variations).

Exploring the Logic of Functions

8. A Visual Demonstration of Development

We begin with the basic pattern of a development of functions, which was introduced previously (diagram 1.04) and is repeated here for easy reference:

A2
A1
A3 1.07

This basic pattern only represents two stages of development, whereas in most cases, the whole process of a development may extend through many stages.

Since the pattern of a development is always the same, the further development of a constituent function, such as function **A2**, can be represented as follows:

A4
A2
A5
A1
A3 1.08

Similarly, if function **A3** develops, the whole arrangement extends as follows:

A4
A2
A5
A1
A6
A3
A7 1.09

If function **A4** develops, the whole arrangement continues to extend as follows:

A8
A4
A9
A2
A5
A1
A6
A3
A7 1.10

If functions **A5**, **A6**, and **A7** also develop, the whole arrangement extends to a complete fourth stage as follows:

A8
A4
A9
A2
A10
A5
A11
A1
A12
A6
A13
A3
A14
A7
A15 1.11

Five stages of development would generate 32 functional variations on the right of the diagram and a total of 63 constituents for all stages:

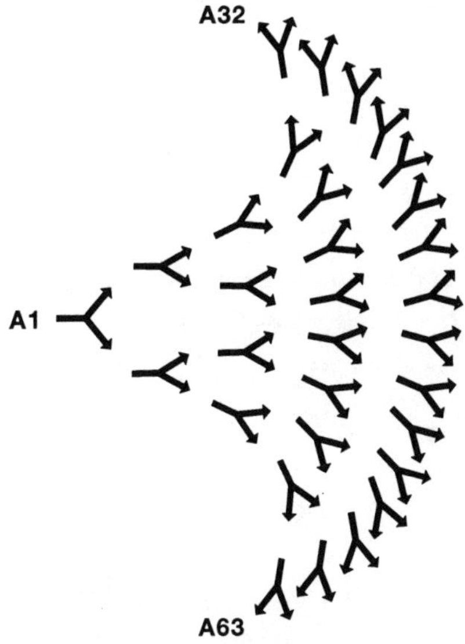

1.12

Thirty stages of development (equivalent to 30 generations of amoebae) would produce over 536 million constituents on the right side and a total of more than one billion for all stages.

9. Looking Toward the Right — the Direction of Development

The progression of diagrams above shows how the basic pattern of development can be used to describe a limited development of functions, an extensive development of functions, or even the whole development of everything. We can see the resemblance between the large form of the most extensive development (diagram 1.12) and the basic form of the original development (diagram 1.07). The scale may change, but the branching form remains constant.

Looking toward the right, we observe that each stage of a development is more complex than the preceding one. The number of functional interactions increases as the number of constituents increases. Since each constituent function determines a pair of functions in the following stage, we can also note, with each successive stage, that the constituent functions diminish in the hierarchy of functions. The development begins with the broadest, deepest function on the left and "ends" with the narrowest, shallowest functions on the right.

The rapid way in which the process of development expands after only a few stages is directly expressed in the development of life on earth. Whenever we feel awed by the enormous numbers of the cosmic bodies, or the vast extent of cosmic distances, we should return from "out there" to contemplate the equally enormous numbers of living cells that have developed "over here."

Every human life begins as a single fertilized egg cell. Within a few weeks, this one cell develops into millions upon millions of living cells, eventually reaching totals that rival cosmic quantities. This gigantic number is again multiplied by the human population of the earth. All the cells of all the other living multicellular creatures — plants, as well as animals — must also be included. Then, by adding the number of living single-cell plants and animals (a number that is probably greater than all the others combined), we come to the total number of the living cells on this earth — a number more awe-inspiring than cosmic numbers. Consider as well that most of these living cells have been generated within our lifetime, many within the last year, whereas the cosmos has taken some 15 billion years to generate. We might reasonably conclude that nothing "out there" supersedes the amazing power of life "over here," neither with respect to its rate of expansion and phenomenal growth, nor to its variety and qualities. The development of life is surely more breathtaking than all we have learned about the development of the cosmos.

10. Looking Toward the Left — an Artificial View

Development proceeds in *one* direction only, toward greater complexity and the generation of more and more variations. Functions only operate in this one direction. This direction is a functional constant. However, our abstract diagrams allow us to examine the form of a development in the "reverse" direction. This view is strictly artificial but very useful. We can see how each constituent function is determined by a specific, deeper function and how each stage of the development reduces, ending on the left with the simplicity of one, single source or function (e.g. the "parent" function **A1** in cell propagation diagram 1.02).

For any given development, the single function on the left is known as the *common functioning principle* (or CFP) of the whole development. This function determines and limits all the functions that develop from it. Hierarchically, it is the deepest, broadest, and most fundamental of all the constituents in the whole arrangement. Every development diagram demonstrates this natural order of functions — the deepest function appears on the left and the narrowest functions appear on the right.

Whenever two functions are related as a pair, the best way to confirm this relationship is to find the deeper function that determines and therefore integrates the relation of the pair. This deeper function, or CFP, must exist because of the basic form of development. However, discovering this function, or any other function in the role of a CFP, is a spontaneous, creative process. We cannot deduce anything that looks backward (i.e. toward the left). Such an action would contradict the direction of development. We cannot *work* backward to the

CFP, even though we can *look* backward in an artificial setting. Everything we do inevitably follows the direction of development, and no action can contradict this functional constant.

11. The Source of Everything

According to the logic of functions, there are profound functions and there are lesser functions. The further we move toward the left of development, the fewer and more profound the functions become. If our development diagram were extended toward the ultimate function on the left, the logic and the order of the development process would reduce all functions to *one*, solitary common functioning principle that determines everything known and unknown. This would be the most profound function, the CFP of all nature, and the function reflective of GOD.

This conclusion is an inevitable consequence of the logic of the development of functions. If the given pattern of development can be shown to be correct and not merely a temporary postulate for a thought experiment, we would have reached a conclusion that penetrates beyond the foundations of contemporary science.

Consider the functional view of the unresolved duality at the heart of classical physics. Physics begins with the function *matter* and the function *energy* — two functions of different kinds that are nevertheless closely related. We might observe that *matter* and *energy* constitute a pair of functional variations and accordingly, recognize that such an association would be determined by a still deeper, unstated function. This description conforms with the basic pattern of a development, and we can diagram the entire relationship as follows:

1.13

In this development diagram, an *unknown* common function [?] determines the two variations, **MATTER** and **ENERGY**. Though **MATTER** and **ENERGY** are different kinds of function, they are nevertheless associated because they have developed from the same, deeper function (represented here as the "unknown" function).

This simple formulation has already led us beyond the artificial constraints of classical science to reveal a deeper realm of function. The "unknown" function of the above development operates in this deeper realm. The logic of development assures us this "unknown" function must exist, and furthermore, it is either one of a pair of *unknown* functions or the ultimate function of nature itself. This is as far as the logic can take us without the further input of new discoveries.

Summary

We began our brief venture into the realm of functional logic with the temporary acceptance of a given set of functional propositions. From that point, we followed the logic of functions, which inexorably led us toward a number of radical conclusions. The given propositions and the various functional conclusions we reached can now be integrated and summarized as follows:

- The given pattern of development is a simple, single-unit form;
- Developments extend by repeating the given pattern;
- Extensions of development based upon the given pattern can account for much of the functional complexity and extent of the known universe;
- The operation of a development establishes a natural hierarchy of functions;

- The whole development of everything begins with a single function — one, solitary, primordial force;
- The functional limit of classical science is expressed by the unresolved duality that constitutes its base;
- There must be at least one function more profound and more fundamental than the functions of *matter* and/or *energy*, which remains unknown to classical science.

Postscript

12. Fragments of Functional Thinking in Our Past

Functional thinking has been practiced and applied within the confines of specific tasks throughout recorded history, but it was never adequately explored or described until this century. Creative inventors, engineers, builders, architects, artists, and researchers must at times think functionally to create products that function. The prehistoric inventors of tools, fire, and the wheel certainly grasped some aspects of the logic of functions.

A column in a building has to be adequately designed to fulfill its whole structural function. It needs to be strong enough to support the live and the dead loads imposed upon it, as well as resist a number of other possible forces. Thinking about the work (i.e. the function) of a column in this way is *functional thinking*. However, those who need to think functionally within a specialized context are frequently unaware of the broader implications of this "technical" way of thinking.

The thought of a single, solitary GOD represents an extraordinary breakthrough of functional thinking. As the Old

Testament reiterates, the *oneness* of GOD is the essential characteristic of GOD. One force determines the whole universe. Here, some 3800 years ago, functional thinking (which includes organ sensations and spontaneous perceptions) led Abram of Ur (later named Abraham) to the same extreme conclusion that we have drawn from a backward view of development (section 11). Right up to the final years of the 19th century, most scientists included Abram's idea of an ultimate function among their basic references.

Another great thinker of the past, Heraclitus (540-470 B.C.), recognized the functional principle of paired, opposite functions. He also derived and promoted the idea that everything is eternally in flux — a thought extending beyond what we can actually observe. We do not know how he arrived at this conclusion, but it is a very profound, functional description of reality.

Finally, we return to modern physics and the search for a "unified field theory." Underpinning this search is a common feeling that the entire operation of nature must be essentially lawful and unified. The most eminent of physicists admit that this idea originates with a feeling. In this one instance, at least, they continue to listen to their feelings. The concept of one GOD expressed this feeling for thousands of years, and now the logic of functions leads us to the same conclusion. Wilhelm Reich not only explored the origin of this feeling within us but also discovered the actual function that generates this feeling — the function that objectively unifies the *whole* of nature.

2

BASIC ORGONOMETRY

Reich's Technique for Comprehensive Thinking

Sections

1. Functioning and Functions
2. Intuitive Beginnings
3. Early Confirmation of Functionalism
4. Relation of Variations to a CFP
5. Derivation of the Basic Symbol
6. The Symbol for Paired Variations
7. Homogeneous and Heterogeneous Variations
8. Operations of Homogeneous Paired Functions
9. Functional Transformations
10. Operations of Heterogeneous Paired Functions
11. The Symbol for Functional Development
12. The Basic Form of a Development Equation
13. The Continuation of a Development Equation
14. The WHOLE Function
15. Transformations in WHOLE Equations
16. Primary Development and the Absence of Zero
17. The Symbol for Fusion
18. The Meaning of Creation
19. The Function of Creation
20. The Expression of Creation
21. Expressing the New Way of Thinking
22. The Integrated WHOLE
23. Nature Integrates the Observer
24. The Role of Orgonotic Contact
25. Directions of Research
26. Directions in Thought
27. Examples of the Two Directions of Research
28. Reconstruction of a Past Development
29. Opposite Concepts: Finity and Infinity
30. Reconciling Conflicting Concepts
31. The Simple and the Complicated
32. No "Cause" or "Effect" in Functioning
33. The Practical Priority of Qualities
34. Quality and Quantity as a Functional Pair
35. The Three Forms of an Orgonometric Equation

It has already become clear, and it will become clearer and clearer, that the old ways of thinking have ended.

WILHELM REICH (1)

In a deep sense, orgonometry is the "language" of the new way of thinking developed by Wilhelm Reich. It is an intellectual tool of enormous potency that can guide us to the resolution of everyday problems, as well as direct the investigations of natural research. Orgonometry is so broad in its scope that we can all benefit from applying it to whatever we do in life.

The regular use of applied orgonometry has led me to many exciting observations I would like to publish. However, all these developments are described in the "language" of orgonometry, and their functional meaning will not be understood without some prior familiarity with the "language" or technique.

Until now, the only reference source for the technique of orgonometry has been the seminal article by Reich, published in 1950. Unfortunately, this article is long out of print and very difficult to find. It will no doubt be republished someday, but meanwhile the rapid growth of my own work has compelled me to undertake the daunting task of formulating a new introduction to the basic technique of orgonometry.

I approached this task with due caution since no one, including myself, can as yet match Reich's mastery of functional thinking. I was astonished to find how the profound objectivity of this abstract system of thought seemed to determine its own form. Sometimes the description appeared to generate itself. At other times, I had to follow Reich's text very closely to maintain the entire functional meaning, specifically where I disagreed with the text and where there seemed to be no alternative way to order the sentence without altering the functional meaning.

The basic content of the following chapter is derived from Reich's "Orgonometric Equations: I. General Form" (see Bibliography). I have reorganized the presentation and created a different sequence; given alternative examples wherever appropriate; completed some explanations that Reich left unstated; and omitted some observations that can be found elsewhere in Reich's published works. All the original equations are represented, but all have been assigned new reference numbers. Three more equations are included: 2.15, 2.20, and 2.23. The first two were abstracted from the written text and the last from a preceding equation.[1]

1. Functioning and Functions

Existence is a state of continuous functioning, and we perceive that everything functions in some way. By "function," we mean *works* or *makes work*.

For example, a *stick* functions as part of the material world. It is similar to a *stone* in that both function as material or inert matter. Although they differ in their origins and other details, they are alike in this one essential way. Now, if we deliberately cut the stick to an exact length, say a yard, we will have altered one of the detailed properties of the stick but not its essential material function. It still remains a *stick*. But when we use it to measure something, the stick acquires a new function, namely, *yardstick*. The *yardstick* does not lose its original

[1] Preface to the first edition of *Basic Orgonometry*, New York, September 1984.

material function, but the new application establishes its new function. However, the function *yardstick* only operates in the context of performing the *work* of measuring. Outside of this activity, the stick resumes its natural function, *stick*. We can see from this example it is the *activity* or *work* that expresses the function.

2. Intuitive Beginnings

Wilhelm Reich tells us that, early in his career, he was biased in favor of certain intuitive premises, and he treated these premises as if they were axioms (2). Gradually, over a period of two decades of original research, each one of these premises proved to be correct. They became part of the basic principles of functional thinking and of orgonometry.

These early premises can be summarized as follows:

1. Energy, in some way, is more fundamental than matter;
2. Antithetical functions are somehow similar and different simultaneously;
3. Direct observation of natural processes is crucial;
4. Human emotional life is accessible to research;
5. A follow-up of 1 and 4 above implies emotional functions must also be essentially an energy expression.

3. Early Confirmation of Functionalism

Reich's initial research work was entirely clinical. Its method relied upon the observation of expressions, that is, the expressions of functions, without the aid of mechanical measuring devices. Nevertheless, he was able to make great strides in understanding the "mechanisms" of neurosis and to uncover the deep, previously inaccessible biological function, namely, the function of the orgasm.

This enormous breakthrough was convincing evidence his functional method of research and reasoning was closer to the operations of nature than the prevailing research techniques. His method demonstrated that biological functions were best described in terms of their *intensities* or *qualities*, which coincides with the way we perceive them. Further, this qualitative approach exposes the relationships between connected functions.

4. Relation of Variations to a CFP

Continued research and repeated observations led him to a fundamental discovery about the relationship of natural functions in the realm of biology.

First, every function is closely associated with one other function. Each function of the associated pair is different, and the difference varies according to the qualities of the associated functions. The difference may be simple, or it may extend to being completely antithetical.

Second, the association of the two functions is determined by a third function, a deeper, broader function that Reich called the *common functioning principle.* This deeper function unifies the two functions that constitute the pair, since both of them develop from this common functioning principle, or CFP. The qualitative characteristics of this deeper function (CFP) determine the characteristics of each of the functions of the associated pair.

Cell division (mitosis) illustrates these basic functional operations. We observe a turgid amoeba **A** (the CFP) under a microscope. After a while, it undergoes a plasmatic convulsion, followed by a complete splitting into two daughter cells, amoebae **A1** and **A2** (the paired functions). Amoeba **A** *develops* into amoebae **A1** and **A2**, and the characteristics of amoeba **A** determine the characteristics of amoebae **A1** and **A2**. The paired functions, amoebae **A1** and **A2**, are related to each other by their common origin (amoeba **A**), yet at the same time, they are independent individuals that are qualitatively *simply* different.

5. Derivation of the Basic Symbol

This basic "one-becomes-two" operation of functions should not be confused with the linear operations of arithmetic or sequential thinking. A diagram can represent it more immediately, more economically, and more accurately than the structure of any sentence that describes it. Reich designed the following diagram to illustrate this universal operation. We recognize it now as the symbol of orgonomy:

Often presented without any labels, this functional diagram represents the role of the *common functioning principle* (**A**) in relation to the paired functions which express it (**A1** and **A2**). We have added the reference labels to emphasize how aptly this diagram describes the development of the amoeba in mitosis. The parent cell is the CFP (**A**), and the daughter cells are the paired functions or variations (**A1** and **A2**).

New facts and a new way of thinking demanded a new technique for formulating and abstracting the operations of natural functions. Reich logically used the above symbol as his model and adapted it to form the basic orgonometric symbol for the *operation of development*:

A ⨎ A1 / A2

This figure closely resembles the orgonomic symbol. By adding the same labels as before, we see that the new arrangement continues to describe our amoeba example correctly. We will discuss the full meaning of this functional operation later. Here we note only that the ƒ across the figure defines it as a *functional* operation.

6. The Symbol for Paired Variations

The *operation of development* symbol, introduced above, shows how paired functions are developed from their *common functioning principle*. It does not provide any information about how the two functions interact with each other. Since such operations are extremely important for understanding the *whole* of a functional process, Reich devised another basic symbol expressly for this purpose:

A1 ƒ A2

This arrangement allows us to consider the relationship of paired functions, like that of the two daughter amoebae, without including the CFP. However, we note the existence of a CFP is implied even though it is not stated, because the two functions paired in this way are identical with respect to their CFP.

As we shall see, the simple form of this symbol accommodates graphic additions that convey further details about the interactions of paired functions. When the setting is both functionally and numerically correct, this symbol can also be modified to represent functional *equality* (3).

7. Homogeneous and Heterogeneous Variations

Functions that are paired because they develop from a *common functioning principle* are not always of the same kind, i.e. *homogeneous*. They may be of a different kind, i.e. *heterogeneous*. Daughter amoebae are of the same kind and are, therefore, *homogeneous* functions. The parent amoeba is also of the same kind, which means that in this example, the CFP and the variations are all *homogeneous* functions. The contraction and expansion of the heart muscle, or the north and south polarity of magnets, are also *homogeneous* functional pairs. Although these paired expressions are antithetical, each pair is the expression of the same kind of functioning process.

The formulation of a *homogeneous* functional pair can be distinguished by its operational symbol and by the identical letter-labels used in the abstract presentation. For example, each amoeba is represented by an **A**.

Functions of different kinds, such as secondary energy and matter, or time and length, are also paired by their common development. However, they do not interact in the same way as *homogeneous* functions. Instead, we find that some *heterogeneous* functions *transform* into one another, and this operation can only occur between functions of different kinds.

The formulation of a *heterogeneous* functional pair can be distinguished by its specific operational symbol, by the different letter-labels used in the abstract presentation (e.g. **x** and **y**), and in some cases, by the two equations required to complete the formulation.

All these distinctions will become clearer after we have examined the complete range of functional operations expressed by paired functions.

8. Operations of Homogeneous Paired Functions

Functions that are paired are always *variations*, since they are never perfectly identical. In most cases, they differ enough to *interact* with each other. We can abstract these interactive operations by simply adding *directions*, in the form of arrowheads, to the basic operational symbol. Reich found there were only four interactive expressions that occur between *homogeneous* pairs of functions in nature.

Simple variations— functions that are alike but nevertheless are also individually unique[2]:

A1 —∫— A2 2.01

The two daughter amoebae, mentioned earlier, are a good example of this relationship. However, any individual amoeba is related to any other individual amoeba in exactly the same way, according to their common function, *amoeba*.

Simple opposites — antithetical functions that attract each other and coexist:

A1 →∫← A2 2.02

Within the realm of sexual differentiation, *male* and *female* variations of any one species attract each other and coexist. The opposite poles of magnets likewise attract each other. *Positive* and *negative* electrically charged particles also demonstrate the same mutual affinity.

Antagonistic opposites— antithetical functions that exclude each other:

A1 ←∫→ A2 2.03

[2] This basic symbol is often used in a less specific way. Therefore, we should not assume the symbol always means "related as *simple variations*."

This "flip-flop" interaction is commonly found operating between the expressions of paired living functions. When an organism expresses *anxiety*, *sexuality* is absent; when it expresses *anger*, *fear* is absent. The functions of the parasympathetic and the sympathetic nerve systems express this antagonistic relationship and, most obviously, where they both supply the same organ (e.g. the heart, bronchi and pupil).

Alternating opposites — antithetical functions that exclude each other but continually alternate:

A1 ⇄ A2 2.04

Regular alternation between exclusive opposite paired functions is expressed as rhythmic motion in living organisms. The heart and the intestines exhibit this motion. Both express the alternate operation of muscular *contraction* and *expansion* in the motion of the heartbeat and in peristalsis. Many manmade devices use this alternating action, for example, the internal combustion engine and long-distance power transmission of electrical energy.

9. Functional Transformations

There are some paired functions in nature that can be transformed into one another. For example, a chemical reaction can be transformed into electricity, and electricity can be transformed into a chemical reaction. We call this operation a *functional transformation.*

Functional transformations also develop from the deeper function, or CFP, toward the variations. We will examine this important operation later, but in all cases, transformation can only occur between *heterogeneous* functions. *Like* functions are already alike. Therefore, as previously noted, each heterogeneous function must be represented by a different abstract letter-label.

Functional transformations between heterogeneous paired functions are also a form of *development*. Therefore, the direction of the transformation must continue to express the direction of development ⟶ toward the right. This means two separate equations are required to describe the two-way transformations of functions that transform into each other.

The symbol for the operation of a transformation between paired functions takes the form of an arrow:

The arrow visually expresses the direction of the functional *transformation*, and it continues to indicate the direction of development. The one-way expression of this symbol distinguishes it from the symbols that describe the operations of homogeneous paired functions (see section 8).

The operation of the functional transformation, as Reich noted, acquires great significance in the investigation of functions that derive from mass-free orgone energy.

10. Operations of Heterogeneous Paired Functions

Functional transformations—heterogeneous functions that transform into each other:

x ⇸ y 2.05

y ⇸ x 2.06

Technological examples of functional transformations have become commonplace in this century. Mechanical energy transforms into electrical energy and electrical energy into mechanical energy. Heat transforms into mechanical motion, and mechanical motion generates heat. Matter partly transforms into nuclear energy, and nuclear energy partly transforms into matter.

Reich used the transformations between energy and matter as his example:

E ⇸ M

M ⇸ E

11. The Symbol for Functional Development

Functional variations develop from a *common functioning principle* (CFP), and this operation is represented by the specific orgonometric symbol introduced briefly in section 6:

⇸<

The arrowheads define the direction of development. The direction is important because it establishes the precise meaning of the symbol. Observe that this symbol can be drawn in either of the following ways:

x, **y** >⇷ **A** or **A** ⇸< **x**, **y**

Both versions show that **A** is the CFP, and functions **x** and **y** develop from function **A**.

In practice, it is convenient to direct the operational symbol toward the right-hand side, thereby placing the CFP, or deeper, broader function, on the left, with the *variations*, or higher, narrower functions, on the right. This means the direction of development → is understood to be moving from left to right.

If we continue this practice, the symbol for the *operation of development* will always look like this:

$$A \prec \begin{matrix} x \\ y \end{matrix}$$

An agreed direction of development helps to distinguish this symbol from its look-alike, the fusion symbol, which will be presented later.

12. The Basic Form of a Development Equation

The basic form of the development of functions can be abstracted as follows:

$$A \prec \begin{matrix} A1 \\ A2 \end{matrix}$$ 2.07

Function **A** dissociates to become a functional pair, **A1** and **A2**. The *common functioning principle* expressed by **A** continues to exist in the functions **A1** and **A2**. This arrangement describes how most natural functions develop. The only exceptions are functions that are *created*, and these will be discussed later.

Returning to our earlier example: When amoeba **A** divides and becomes amoeba **A1** and amoeba **A2**, the qualities of amoeba **A** are not lost but continue to be expressed by amoeba **A1** and/or by amoeba **A2**. Each of the two daughter amoebae has essentially the same qualities as the parent amoeba **A**. Thus **A** is functionally *homogeneous* with **A1** and **A2**, as indicated by the common letter-label **A**.

We note that **A** is not the sum of **A1** and **A2**, nor is **A1** or **A2** equivalent to half of **A**. Numerically, cell division produces a strange condition where one unit transforms into two, and two transforms into four, and so on. Thus, they cannot be equated directly. We must admit that quantitative thinking does not lead to an adequate understanding of the relationship of functions.

Heterogeneous functions require different letter-labels for each different function. A development equation with heterogeneous functions also expresses the qualitative *transformations* that occur in the development:

$$\mathbf{A} \prec \begin{matrix} \mathbf{x} \\ \mathbf{y} \end{matrix}$$ 2.08

This equation shows the development of a transformation from the CFP toward the variations. The CFP, **A**, transforms into function **x**, or function **y**, or both. To comprehend the entire transformation process, we would also need to know how the paired variations, **x** and **y**, interact (see section 10).

13. The Continuation of a Development Equation

The continuation of the functioning principle throughout all the dissociations of development can be formulated in this way:

$$\mathbf{A1} \prec \begin{matrix} \mathbf{A2} \prec \begin{matrix} \mathbf{A4} \\ \mathbf{A5} \end{matrix} \\ \mathbf{A3} \prec \begin{matrix} \mathbf{A6} \\ \mathbf{A7} \end{matrix} \end{matrix} \quad \longrightarrow \textbf{etc.}$$ 2.09

The functioning principle of **A** is continued in **A1** and **A2**. **A1** and **A2** form a functional pair, since they are identical in relation to their CFP, function **A**. **A3** and **A4** develop from **A1**, and they also form a functional pair identical with regard to their CFP, function **A1**. Similarly, **A2** is the CFP of the functional pair, **A5** and **A6**, and so on.

Cell propagation begins with amoeba **A**, and after two stages of division, we have four amoebae, **A3**, **A4**, **A5**, and **A6**. Each one of the four cells continues the functioning principle derived from amoeba **A**. All are identified by their common origin, and all are simultaneously different or individually unique.

A development equation is integrated by the functional characteristics of its CFP. These characteristics pervade all the constituents of each stage of the whole development. It is not the differences of function, but the *functional identity* that determines the integration of the constituents, and that unifies the whole expression of a realm of nature, or of a correctly formulated description of nature.

14. The WHOLE Function

Natural systems and manmade systems function as a *whole*, and a system's function, or *whole* function, is determined by the common functioning principle that pervades all of its parts. The *quantity* of the parts that constitute a system has no relevance in determining the *whole* function. For example, one brand of camera is constructed of 160 separate parts, whereas another consists of 240 parts. Both have the same whole function, namely, *camera*. The parts of each brand are integrated by this function irrespective of their number. Thus, variations in the number of parts, speeds, energy consumption, and so on, do not affect the CFP as long as the specific *whole* exists, and the *whole* function continues as long as the CFP remains the same.

However, if the CFP changes, then the *whole* function will alter too. Consider the case of a factory that formerly manufactured *cameras* but is now making *projectors*. The building and the machinery remain the same, but the CFP has altered. Functionally, the factory now constitutes an entirely different *whole*. In time, as the operation of this new *whole* function continues its development, technical adjustments will alter not only the original machinery, but also the building. The factory workers are governed by the same CFP, and they, too, are altered according to the work they do.

15. Transformations in WHOLE Equations

Transformations of a *whole* function operate from the CFP toward the *variations* and between the *variations*. The crucial function that constitutes the change of the CFP is to be found among them. The following three *operation of development* equations illustrate, in the abstract, the different kinds of transformations and interactions among the constituents of the *whole* function:

x1
x
x2
A
y1
y
y2 2.10a

x1
x
y1
A
x2
y
y2 2.10b

a
x
b
A
c
y
d 2.10c

Each equation represents an entirely different *whole* function. The constituents of equations 2.10a and 2.10b remain the same, but the types of transformation and, therefore, the details of the integration are different. The constituents of equation 2.10c are all qualitatively different, and this formulation represents the most complex development of transformation operations.

The development of variations, together with the operation of transformation, is a continuing natural process, and the possibilities are endless. However, at the root of it all, there is only *one* single and final CFP that determines everything.

16. Primary Development and the Absence of Zero

The basic form of the *development* of functions and the arrangement of *whole* development equations express the direction of development ⟶ toward the right. If, however, we follow all existing CFPs toward the utmost left ⟵ of the equations, we reach a solitary CFP and a number *one*. The logic of functions leads us to this conclusion, and Reich's scientific discoveries confirm this is indeed true.

The prevailing concept of an *absolute zero* appears to contradict the above conclusion, but we cannot find *zero* anywhere in nature. *Zero* simply *does not exist*, except as a creation of thought. Experiments have confirmed there is no absolute vacuum in space, and space is filled with orgone energy as far as it can be determined (4).

The one and only primordial CFP unifies all existence, pervades every developed function and every realm of functioning, and is expressed throughout all of nature. It is the CFP of *all* nature, and its *primary development* can be abstracted and generalized as follows:

$$N \text{ ⤙ } \begin{matrix} Vx \\ Vy \end{matrix}$$ 2.11

where **N** represents the CFP of all nature, **x** and **y** represent the first two variations, and **V** represents the principle of variation. The equation states that the primordial function **N** develops the paired variations, **Vx** and **Vy**.

Each single variation on the right of the equation will vary in accord with the primordial function of nature; each variation will function as a variation determined by its CFP on the left; and each will simultaneously function as a CFP for its developed functions on the right.

17. The Symbol for Fusion

The symbol for the *operation of fusion* was evolved to describe the specific transformation that occurs in the moment of creation. Briefly, it represents a functional operation wherein two primary variations superimpose and *fuse* to create a single *new unit* of function. The fusion of a male and female germ cell, which creates the fertilized egg cell, is a typical example of this process.

If we maintain the direction of development ➞ as a movement toward the right, the symbol for the *operation of fusion* will appear to be the reverse of the previous symbol for the *operation of development*, except for the arrowhead. The arrowhead will continue to point in the direction of *development*, and the symbol takes the following form:

$$\begin{matrix} Vx \\ Vy \end{matrix} \text{ ⤚ } A1$$

The symbol illustrates the *operation of fusion*. The primary variations, **Vx** and **Vy**, *fuse* to create a single, new function, **A1**.

As we have indicated, this operation is also a form of *development*. But, since it is contrary to the basic operation of development, we exclusively refer to it as the *fusion* or *creation* operation.

18. The Meaning of Creation

In the context of orgonometry, *creation* specifically means the development of:

- New individual units within the process of a whole development, for example, the offspring of plants and animals;
- Entirely new forms of existence, for example, the spontaneous development of entirely new species of organism or of new particles of matter.

Both cases express the emergence of a qualitatively different CFP that will now continue to develop its own variations.

Procreation in metazoa does not follow the basic *operation of development* as it does in the case of the amoeba. Highly developed animals do not divide or dissociate in the process of reproduction. The operation appears to be quite the reverse. Two organisms, male and female, make contact, superimpose, and their respective germ cells interpenetrate and fuse. The *fusion* operation creates the new functional unit, namely, the fertilized egg cell, and this cell initiates the development of the offspring. In place of dissociation, *creation* functions through the operation of *fusion*.

Biological creation also occurs in protozoa. Amoebae, for example, sometimes *fuse* and, thereby, create a new individual unit out of two variations. This new unit, or amoeba, continues to propagate itself by the regular process of division discussed previously.

The process of *creation* also functions in non-living nature. Every *thing* in the cosmos was initially created. As Reich observed, "We must assume that the process of *creation* mirrors some basic natural law" (5).

19. The Function of Creation

Functional equations that describe development must unequivocally express the direction of development without relying upon any presumption of an understanding. The following preliminary formulation of the creation function is unresolved about this expression:

$$\begin{matrix} \mathbf{Vx} \\ \mathbf{Vy} \end{matrix} \succ\!\!\rightarrow \mathbf{A1}$$

In *creation*, the direction of development leads away from the CFP (**Vx** and **Vy**) toward a new CFP (**A1**). It differs from the general operation of development precisely because *both* primary variations, **Vx** and **Vy**, constitute the CFP of the singular development **A1**, and because **A1** is also the CFP of future developments. Thus, the distinctive symbol for the fusion operation does not convey, by itself, the *whole* function of creation.

Reich derived the function of creation by integrating two previous equations, equation 2.11 (see section 16) and equation 2.10c (see section 15):

$$\mathbf{N} \rightarrow\!\!\prec \begin{matrix} \mathbf{Vx} \\ \mathbf{Vy} \end{matrix} \succ\!\!\rightarrow \mathbf{A} \rightarrow\!\!\prec \begin{matrix} \mathbf{x} \rightarrow\!\!\prec \begin{matrix} \mathbf{a} \\ \mathbf{b} \end{matrix} \\ \mathbf{y} \rightarrow\!\!\prec \begin{matrix} \mathbf{c} \\ \mathbf{d} \end{matrix} \end{matrix} \qquad 2.12$$

From this, he extracted the basic expression of the *function of creation*:

$$\mathbf{N} \nrightarrow\!\!< \begin{matrix} \mathbf{Vx} \\ \mathbf{Vy} \end{matrix} >\!\!\nrightarrow \mathbf{A1} \tag{2.13}$$

where **N** represents the CFP of all nature (the basic orgonomic natural law), **x** and **y** the concrete variations, **V** the principle of variation, and **A1** the new CFP which emerges from the fusion of **x** and **y**.

The operations in this equation clearly convey the direction of *development*, and its form expresses the singular quality of the transformation in *creation*.

20. The Expression of Creation

It should be understood that the *whole* function of creation is a process of a purely orgone-*physical* nature, that is, a primary functional process which should not be confused with secondary functions, such as mechanical energy or inert matter. The symbol **N** represents the primordial function of *undifferentiated* cosmic orgone energy, and **Vx** and **Vy** the primary differentiations of orgone energy. Only the new functional unit, or CFP, represented by **A1**, manifests the characteristics of the familiar secondary realm of functions.

Reich noted that creation "seems to be limited to free, primordial orgone energy functions only," and the CFP of the whole equation demonstrates this too (6). Nevertheless, this functional process is expressed both in non-living and in living nature. In the cosmos, orgone energy continues to be free and unbounded, and cosmic creation must be continuing discretely. In living organisms, creation appears to be a direct continuation of the primordial, mass-free orgone energy function operating within the membranes of living matter. Mankind expresses this function in common with cosmic functioning. The operations of superimposition, fusion, and creation root us firmly in the midst of the universe. Indeed, all living forms are similarly rooted in the universe (7).

Organismic orgone within cell membranes may be *differentiated*, but it is not structuralized. We recall that Reich's Experiment XX demonstrated that membranous matter itself emerges from the structuralization of free, primordial orgone energy (8, 9).

Creation in living matter is an *irreversible* transformation. **Vx** and **Vy** cannot be reconstituted from the new unit **A1**. Whether this characteristic applies to *all* expressions of the function of creation is not yet known. But my own observations, to be presented elsewhere, strongly suggest that this is an *inherent* characteristic of the function of creation.

21. Expressing the New Way of Thinking

The development of orgonometry began with the spontaneous invention of the basic form of the functional equation. By the time Reich came to write the introductory article "Orgonometric Equations: I. General Form," he had extensively developed this new subject, both theoretically and in practice. He, therefore, included a description of some of these developments along with the presentation of the basic technical material. I shall follow his example and restate all of his observations and conclusions, taking great care to reproduce their functional content accurately.

Some of the thought content in the following sections may be familiar to those who have studied Reich's writings on the functional method of thought (see Bibliography). Indeed, orgonomic functionalism and orgonometry are related. They constitute a functional pair, identical with respect to Reich's perception of natural functioning, as determined by his general functioning during the period when he created these new variations.[3]

Although orgonomic functionalism was developed in association with verbal expression, its objective simplicity is obscured by the structuralized state of our language. The expression of orgonometry is free of this problem, since it is completely integrated with the functions and processes it represents. It provides a more lucid and a more precise way of describing functional concepts. Of the two variations, orgonometry has a far greater potential for future development. As Reich anticipated, it emerges as the dominant method for expressing functional thoughts.[4]

22. The Integrated WHOLE

The constituents of an *integrated whole* function can change and yet continue to be an integrated functional *whole*. The functional concept of the "integrated whole" does not describe a rigid state but a *process*.

To comprehend the whole *as a process*, we need to know more than the present state of the functions. We must also know how they interact and, specifically, their transformations. Generally, we try to keep the concept of the integrated *whole* in mind without isolating any one functional constituent for too long, otherwise we could deviate from the functional approach. In practice and for a short time, we may isolate a *whole* that is not functionally integrated, such as the *compressive strength* of a steel girder. Later, when this girder becomes

[3] This observation is reviewed in chapter fifteen (15.04).

[4] We deduce this from Reich's repeated use of "orgonometry" as a heading for all five articles on "Orgonomic Functionalism," *OEB* 1950-52 (see Bibliography). It is also implied by his statement of future intent that footnotes the first page of two of these articles.

a constituent of an integrated *whole* (e.g. a building frame), it may well function as a *tension* constituent, which is quite the opposite of the function we previously isolated. We see from this mechanical example that the constituent of an *integrated whole* functions according to its role in the *whole* process, and that an isolated characteristic may not describe this functional role.

Nature is much more fluid than any mechanical device, and because it is so fluid, we need to describe the functional transformations precisely and to know the *kind* of integration of the *whole*. Each functional unit must be concretely defined, together with all its transformations. The formulas presented earlier make this task much easier (see sections 2.13 and 2.15). Preliminary formulations of the functions being studied will often help identify what functional information must still be found. The formulations can also focus our attention on otherwise hidden errors of concept.

23. Nature Integrates the Observer

The operations of thought are in some ways removed from the primary operations of nature. Pure abstractions of the mind are of little use in directing the study of functions.

Common experience has shown that abstract reasoning, divested of the input of our senses, does not lead to an understanding of the interrelations of the many unconnected single events we observe in nature. Such operations of thought are too unlike the basic functioning processes of nature to reflect them precisely. To bridge this gap, we first need to comprehend the operation of the common functioning principle in nature. The observer, who is a part of nature, and the realm of nature being observed can be coordinated in the following *real* arrangement:

2.14

The **OBSERVER** and the **OBSERVED** are variations of their CFP, **NATURE**, or cosmic orgone, which governs them both.

According to this functional setting, *nature* integrates the *observer* and the *observed.* Therefore, every orgonometric abstraction must correspond to a concrete, observable, and verifiable function in *nature.* The system of thought should not float *above* reality, that is, disconnected from *nature.* It, too, must follow the same basic laws of *nature.* We discern that the *thought* process and the *objective* process, likewise, constitute a functional pair with a CFP outside of their domain. This CFP is the fundamental constituent of the whole process of comprehensive thinking.

24. The Role of Orgonotic Contact

The process of investigation and the process under investigation form a functional pair governed by a deeper function, or CFP. Reich uncovered this crucial factor and called it "orgonotic contact," meaning the contact between the continued primary functions within the researcher (organismic orgone) and the primary orgone functions of nature (atmospheric and cosmic orgone). The paired functions cannot be directly equated, since their identity is determined by the CFP, according to the following arrangement:

2.15

The CFP limits its own development, thus the *quality* of the **ORGONOTIC CONTACT** will determine the *quality* of the research and, in some cases, the *quality* of the objective process itself.

A researcher who is partly in contact with his own organ sensations and, therefore, limited in his *orgonotic contact* in general, will be limited in his work to a narrow field of study. If, at the same time, the process being studied is itself *responsive*, then the limitation of the researcher could modify the process being investigated. This variation of development is frequently encountered in biological research, and it usually leads to functionally false conclusions.

25. Directions of Research

Scientific research proceeds in one of two directions:

- It investigates the characteristics of a selected functional realm and its whole *development*;
- It investigates the characteristics of a selected functional realm with the intent of finding its deeper origins, that is, its CFP.

In the first case, the research begins with a CFP, **A**, and proceeds toward the variations, **x** and **y**. We can represent this process with the following form:

$$\mathbf{A} \nrightarrow\!\!< \begin{matrix} \mathbf{x} \\ \mathbf{y} \end{matrix} \qquad 2.16$$

We observe that the direction of the *research* proceeds in the same direction as the *development* of the functions under investigation.

In the second case, the research begins with a functional variation, **x** or **y**, and then tries to find its paired function, **y** or **x**, and their characteristic interactions. After this, the search turns to finding their CFP, **A**. We can represent this process with the following form:

$$\begin{matrix} x \\ y \end{matrix} \succ\!\!\nrightarrow A \qquad 2.17$$

The equation shows that the direction of the *research* does not proceed in the same direction as the *development* of the functions under investigation. The two directions are entirely opposite. Thus, in the above representation, the CFP to be discovered appears on the right, and the usual form of the development equation appears to be reversed. (Note this operation symbol resembles the *fusion* symbol introduced in section 17.)[5] However, the reconstruction of the objective conclusion of this procedure would be presented in the usual form, with **A** on the left, and **x** and **y** on the right.

26. Directions in Thought

The two directions of research reflect, in a practical application, a basic characteristic of the *operations of thought*. The operations of thought can move in either direction, while the development of the thoughts ⟶ continues to fulfill the one direction. The preceding equation 2.17 expresses the *development* of thoughts or findings in an investigation that is directed toward the CFP of nature.

Since orgonometry is essentially a technique of *thought*, it entirely expresses the *operations of thought*, as well as the *development of natural functions* upon which it is founded. Thus, there are two directions to be considered in an orgonometric equation. One leads from the CFP toward the variations, namely, *development* ⟶, and this expresses the development of *thought* and *nature* in general. The other leads from the variations toward the CFP, namely, *structuralized,*

[5] The meaning of this operation will be reviewed in chapter eight.

past development ⟵, and this exclusively expresses the operations of *thought* and manmade devices.

The two basic directions in orgonometry correspond to the concept of the "vector" in applied mathematics. But the *vector* only represents a change in quantities, whereas the orgonometric direction concerns changes in *quality*, as well as changes in quantity.

27. Examples of the Two Directions of Research

Reich's original investigations of the development of neurosis and character structure are replete with examples of the two directions of research. They can be studied in great detail from his books and papers. The following examples can only outline the procedure.

Beginning with the neurotic symptom as a CFP, the investigation can proceed toward the consequences of the symptom. Daydreaming, for example, may lead to an accident, a lack of concentration, problems at work, and so on.

If the symptoms are viewed as *variations*, the investigation can proceed backward into the structuralized history of the person. The symptoms are inevitably expressed in pairs: genital anxiety pairs with sexual hyperagility; hyperorderliness pairs with utter disorderliness; and so on. The investigation can then uncover the CFPs of the paired symptoms, sexual stasis and sexual repression. They, in turn, may be paired as variations of a still deeper CFP, orgastic anxiety.

Character analysis proceeds in this way, toward the CFP on the left, until it reaches the bioenergetic core of *all* the neurotic symptoms on the right. But this, too, is a variation. Further investigation will reveal its paired function and a deeper CFP that unites them.

28. Reconstruction of a Past Development

Once the original research is done, the entire biopathic process can be theoretically reconstructed as an orgonometric formula that will show the relative depths and interrelations of the different functions:

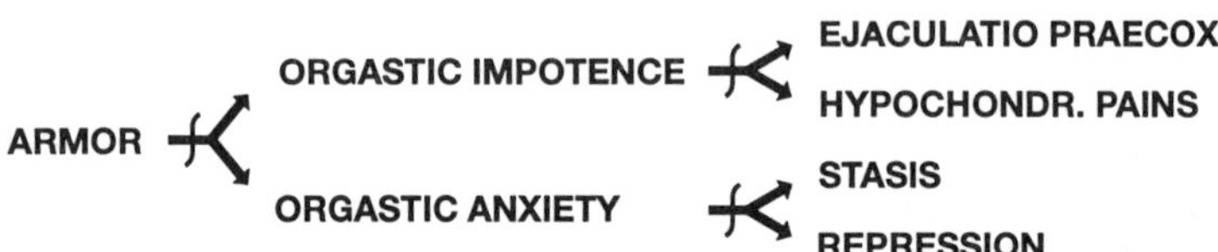

2.18

The CFP, **ARMOR**, is expressed in all the constituents to the right. From left to right, we can follow the whole biopathic development that is governed by the **ARMOR**. When abstracted, the form of this equation shows the type of whole *integration* involved:

A x y a b c d

Compare this form with the others (see equations 2.10a and 2.10b in section 15).

The entire process of biopathic armoring is *reversible*. The armor can be dissolved, and all the constituents will disappear. Most mechanical systems can also be reversed by rearranging their constituents. However, some important developmental processes are *irreversible*, such as living organisms, a valid system of thought, and the *function of creation* (see section 20).

29. Opposite Concepts: Finity and Infinity

The general form of a *development* equation, such as the one above, has two characteristics that relate as a functional pair. The CFP on the left is *finite* and singular (**1**), and the variations on the right are numerous and infinite (∞). The interaction between these two concepts is not immediately clear, so we present them in the preliminary abstract form of a simple variation equation (see section 9):

1 ⨍ ∞ 2.19

The development of variations is an infinite process within any functional realm. Only death or the cessation of the whole functional realm can intervene. *Motion*, for example, develops an infinite variety of paired expressions.

However, the expressions of each variation are limited by the characteristics of its CFP. The CFP pervades and governs *all* the variations that develop from it. Thus, the *freedom* of any development is contained within the limit *determined* by the characteristics of the CFP. *Freedom* and *determinism* are thereby paired, according to the abstract form of the above equation, and can be formulated as follows:

DETERMINISM ⨍ FREEDOM 2.20

We understand from this preliminary setting that **FREEDOM** coexists with **DETERMINISM** in the operations of nature. Further, that *natural law* is paired with an operation that appears to be *lawless*. Natural development does not proceed unless all the constituents of a given functional realm are perfectly fluid and alterable. Development is blocked by rigid, unalterable structures. The structuralization of the exoskeleton of insects, for example, demonstrates dramatically how growth ends with structure. The same principle operates in society, where the established structure denies new concepts and neglects new creations.

The transformation of *development* into rigid *structure* begins when its free movement starts to freeze. *Structure is frozen movement.* We can reconstruct the development that preceded a structure from the evidence of the frozen motion expressed in the details of the form of the structure. This technique is used to uncover the past development of human character structure, to analyze archaeological fossil remains, and to interpret ancient cultures from the ruins of their architecture. We observe, in this context, that *development* and *structure* form a functional pair identified by the CFP, *history*:

2.21

30. Reconciling Conflicting Concepts

The variations discussed above can be grouped into two related sets. The one set includes development, freedom, infinity, and indeterminism. The other set includes structure, law, finity, and determinism. We may personally prefer one set above the other, but natural functioning does not follow characterological preferences. It simply functions, expressing itself through the simultaneous actions of such opposites.

By aligning our thoughts with the actual functioning processes of nature, we find that the two sides of political expression, liberal and conservative, are simultaneously functional. Both have their place in the development of events. Law and freedom, determinism and indeterminism, finity and infinity, are no longer absolute opposites, that is, they are neither irreconcilable nor mutually exclusive. We see them now as inseparable paired functions of opposite qualities that simultaneously express one, fluid, common operating process. This is not a matter of "compromise" or "adjustment of differences," but the *operation of natural law itself.*

This means that conflict, or the coexistence of opposites, is a natural process and not a matter of preference. It also means that sharp contradictions can only be solved *outside* of their domain through knowledge of the action of the "third force," or common functioning principle. Exaggerated thoughts of absolute opposites dissolve when submitted to the functional technique of solving such problems by means of the CFP. The conflicts between the individual and society, between religion and sex, between mechanism and mysticism, are but a few Reich was able to resolve by applying this technique.

31. The Simple and the Complicated

The orgonometric form of a *development* equation shows, as we follow the direction ⟵ toward the CFP of all nature, that each CFP along the way represents a wider, deeper realm of functioning in which the interrelation of functions is progressively simpler. The CFP of the phylum *pisces* is much wider and simpler than the varied constituents and behavior of any single, individual fish. The opposite direction ⟶ moves through a progression of narrower, higher domains of function toward an ever-increasing complexity of more and more variations.

32. No "Cause" or "Effect" in Functioning

The concept "cause" and the concept "effect" do not appear in any orgonometric equation. Amoeba **A** does not "cause" amoeba **A1** and **A2**, and the latter are not the "effects" of amoeba **A**. If we are to use these terms, we would have to say **A1** and **A2** are the "effects," or "results," of the process of cell division but not of the function, amoeba **A**.

Viewed as a CFP, a function *develops, determines, limits, pervades,* and *integrates* its variations, but it does not "cause" anything. However, in the course of a functional development, "cause" and/or "purpose," and "effects" and/or "results" can be abstracted. *Purposes are derived from functions* and *effects from processes*. In nature and in orgonometry, there is no direct relationship between "cause" and "effect," or between "purpose" and "results."

For example, if we decide to fly to New York (i.e. our purpose), the function *airplane* would have been developed and existing prior to our decision. In turn, the function *airplane* was not invented for this specific purpose, but was developed out of previously invented functions, such as the internal combustion engine, light-weight frame structures, and the air propeller. The inventor of the first internal combustion engine did not know beforehand how his engine would be applied in the future. We see from this example that *the function always precedes its purpose*. If we are to align our thinking with the functioning of nature, we must recognize "purpose," "use," and "goals" are secondary to the function from which they are derived.

Functional thinking discards or replaces many familiar mechanistic concepts. The CFP replaces "cause"; variations replace "effects"; the integrated whole, or the functional unit, replaces the "sum total"; and functional constituents replace the "parts."

33. The Practical Priority of Qualities

Every natural function has two basic characteristics:

- It expresses certain *qualities*, such as form of motion, color, shape, density, viscosity, consistency, and so on;
- It expresses *quantities* in association with the qualities, such as speed in motion, rates of change, weight, length, volume, rotation, and so on.

In our contact with nature, we immediately respond to its functional properties and not to its quantities. The quantities and the measuring techniques we apply are artificial interferences created and developed by the human intellect. Nature continues to function and we try to measure its functioning. Our appreciation of function is itself a function of perception and this must be taken into account. The immediate perception of any function is entirely qualitative, and in practice, this determines that the *qualities* of a function take priority over the *quantities* of a function.

Engineers constantly experience the practical priority of functional qualities. For example, before a structural engineer can analyze and calculate a reinforced concrete beam, he needs to know all the relevant functional qualities of the beam, such as its shape, disposition, kinds of support, and so on. This means that the general idea of the beam, of the form of the rest of the structure, and of the architectural function as a whole must be there first.[6]

Certain types of investigation commonly commence with measurements, even though the function being measured is not properly understood. Statistical surveys are repeatedly begun in this way with the false hope of reaching a deeper

[6] The work of the architect, which precedes that of the structural engineer, requires a deep understanding of the functions of material and of structure.

understanding. As we have already shown, this is not possible, since we first require to understand *what* reacts and *why* it reacts *before* any meaningful measurements can be made. The faith we have invested in numbers and mathematical computations is not well founded. Reich observed that because numbers and equations are resolved to zero, they invoke the illusion of a "perfect" state. But nature is not perfect in this way, and we, as a constituent of nature, cannot expect to be perfect either.

A functional investigation commences with a qualitative description of the functions being studied. By linking up the observed functions according to their *qualitative* expressions, we can come to understand *how* the whole functional process develops. We first set out to understand the whole function as completely as possible. We do not restrain our formulations with numbers until we comprehend the *whole* function. This is also the way to approach the formulation of *complete orgonometric equations* which include both the qualities and the quantities of the functions.

34. Quality and Quantity as a Functional Pair

Quality and quantity do not equate directly, but they are identical with respect to their CFP, that is, they are both properties of functions. We can coordinate them in the following setting:

2.22

The equation shows that **FUNCTIONAL PROPERTIES** are simultaneously expressed through **QUALITY** and through **QUANTITY**.

The *qualities* are immediately perceived (*subjective*), whereas the *quantities* are conceived and measured (*objective*). Qualities have *intensity*, and quantities have *extensity*.

When Reich measured the bioelectrical effects of pleasure and anxiety (1936), he found the *intensity* of the emotion was functionally identical with the *quantity* or *extent* of the bioelectrical charge at the skin surface (10). He thereby established, for the first time, that the observer and the observed, the subjective and the objective, quality and quantity, intensity and extensity, are paired constituents of a single whole functioning process.

The priority we assign to the qualitative functional formulation follows logically from the requirement that functions must be understood before we try to measure them. This approach is essential if we are to arrive at *complete* orgonometric formulations. The above equation shows that a *complete* orgonometric equation would include both qualitative and quantitative expressions.

In his second paper on orgonometry, "Complete Orgonometric Equations," Reich demonstrates how the quantitative definitions evolve spontaneously out of the functional definition, when the qualitative description is correct (see Bibliography).

35. The Three Forms of an Orgonometric Equation

First we observe how anything functions by responding to the expressed *qualities*. Then we relate the constituent functions according to their qualities. Those which do relate can be arranged in one of three, simple functional patterns, and from these we can abstract the three basic forms of an orgonometric equation.

The form of paired functions:

$$\mathbf{Vx} \not\!- \mathbf{Vy}$$ 2.23

where **Vx** and **Vy** represent variations of function that are different in a simple or extreme way yet, at the same time, are unified by a common third function, or CFP, not included in this basic form.

The form of the development of functions:

$$\mathbf{N} \not\!\prec \begin{matrix} \mathbf{Vx} \\ \mathbf{Vy} \end{matrix}$$ 2.24

where **Vx** and **Vy** represent variations as before, and **N** represents the common third function, or CFP, which was absent in the previous form. The function **N** both determines and unifies the variations **Vx** and **Vy**.

The form of the function of creation:

$$\mathbf{N} \not\!\prec \begin{matrix} \mathbf{Vx} \\ \mathbf{Vy} \end{matrix} \succ\!\!\not\!\rightarrow \mathbf{A1}$$ 2.25

where **N** represents the primordial CFP, i.e. the primary function of all nature, **Vx** and **Vy** the primary variations, and **A1** the new unit created by the *fusion* of two primary variations. The right half of this equation symbolizes the operation of *fusion*, which only occurs in the act of creation.

The fundamental simplicity of orgonometric formulations is a direct reflection of the functional simplicity at the root of nature. Three simple arrangements are sufficient to describe the infinite variety we encounter in nature.

3

THE REAL MEANING OF $E = mc^2$

Sections

Purely logically, it is improbable that two so different entities as energy and mass should be simultaneously primary, and classical physics, including the modern energy-mass relation, conceived of both mass and energy as primordial natural phenomena.

WILHELM REICH (1)

The presentation of this chapter closely follows the text of my first orgonometric paper on a scientific subject. The whole subject could be condensed into a page or two of orgonometric formulations with a minimum of word descriptions. However, promoting functional thinking and promoting the use of orgonometry by means of practical examples is also an important part of my work. When I first wrote the paper, I thought of it not only as a subject but also as a practical example of the application of the technique and, therefore, I included an account of the thought processes and the orgonometric thought experiments involved in the original investigation.

The subject originated with a simple, but significant observation about the functional content of Einstein's famous equation for energy and mass. I recognized this equation was a good example of a *quantitative* orgonometric equation for paired variations. When the functional content of Einstein's equation is restated as *qualities* rather than as *quantities*, the reconstructed formula raises some interesting questions about its *whole* functional meaning.

After exposing the mechanistic limitation of the original formula, the chapter begins again with a purely orgonometric demonstration. Even without reference to Einstein's equation, the logic of functions leads us to the same set of questions: The meaning and depth of function E; the operation of the known transformations; how to integrate these transformations; and the role of the unstated CFP. The functional answers generate a new pair of qualitative equations more comprehensive than the accepted meaning of $E = mc^2$. Some extra notes are presented as an addendum.

1. The Functional Content of $E = mc^2$

The close relation between matter and energy was defined by Albert Einstein in 1905 with the following formula:

$$\mathbf{E = mc^2} \quad 3.01$$

It tells us that a quantity of *energy* (**E**) equals a quantity of *matter* (**mc^2**) measured as mass (**m**) times the speed of light squared (**c^2**). According to the numbers, we can expect to find that even a tiny quantity of *matter* equals an enormous quantity of *energy*. This was confirmed dramatically and destructively when the first atomic bombs were dropped at the end of World War II.

Though formula 3.01 looks simple and basic, it is neither. Wilhelm Reich includes it among his examples of non-basic formulas, which express a mixture of functions in an association that is functionally meaningless or "false" (2, 3).

Nevertheless, formula 3.01 must be correct in some way. It could not be used successfully if it were not. Let us therefore extract the functional content of the formula and restate it as a word-labeled functional equation:

ENERGY ⊹ MATTER 3.02

This translation presents the function **ENERGY** and the function **MATTER** as a pair of related variations. It states that **ENERGY** and **MATTER** are simultaneously *identical* and *anti-*

thetical with respect to an unstated *common functioning principle* (CFP).

We can immediately observe a serious functional inconsistency expressed by this orgonometric statement. According to orgonomic usage, the function *energy* specifically represents *primordial energy*. As such, it cannot be paired with other functions. The function *matter*, on the other hand, expresses the *narrower* realm of secondary function, and it can only be paired with another secondary realm function.

If this were not the case and we were to accept the given relationship as stated, we would still have to account for the unstated CFP that determines the relationship of the given functional pair. This unstated function has to be deeper and broader than either of the paired functions, and this fact would show us that the given functions are not really fundamental in the first place.

We are forced to conclude that "energy," the given term, is intended to represent the function of *secondary energy*. This realization immediately clarifies the realm within which Einstein's equation applies. We can now adjust and restate our qualitative word-labeled version of his formula as follows:

SECONDARY ENERGY ⨍ MATTER 3.03

This equation states that **SECONDARY ENERGY** and **MATTER** constitute a related pair of functional variations. They are identical with respect to a deeper, *common functioning principle* understood but not specified in this form of equation.

Equation 3.03 accurately reproduces the functional and qualitative content comprehended by Einstein's algebraic equation. However, this formulation also continues to show that the functional content of the original is neither basic nor complete.

First, we are reminded of the missing CFP, the deeper level of functioning, which is not even implied by the algebraic original (formula 3.01). Second, it raises a crucial question about relating two different kinds of function, namely, *heterogeneous* functions — a distinction technically disregarded by the method of algebra. Thus, the orgonometric formulation highlights what is qualitatively unresolved or *pending* in the operation described by the original equation.

To complete the orgonometric representation of the above operation, we need to know the precise way in which the two functions interact. Since the functions are obviously of a different kind (heterogeneous), we would want to know whether, in some way, the two functions do or can transform from one into the other. Einstein's formula does not convey any information about such operations, but a physicist would understand such a conversion or transmutation could theoretically occur.

We pause to observe how the abstract operation of algebraic equality *floats above reality*. It does not necessarily represent a real exchange or a factual transmutation. It is only real with respect to the abstraction of numbers.

The general belief that Einstein's formula is a very basic description of nature exemplifies the limitations of the reasoning methods used in physics and mathematics. From the viewpoint of orgonometry, this formula is no nearer to being a primary description than any other formula of classical science.

2. Investigating a Perceived Functional Relationship

The technique of orgonometry can lead us directly and more comprehensively to the same functional formulation as expressed by equation 3.03 without any reference to Einstein's equation. To prove this, and as a practical demonstration of applied orgonometry, we will now present an orgonometric investigation of the perceived relationship between "energy" and "matter."

We begin with the perception that *matter* and *energy* are somehow related. Our first step is to clarify the basic orgonometric form of this operation — the pattern of the relationship — so we consider the following questions:

- Do the two functions constitute a functional pair, or are they somehow more indirectly related?
- Does any one of the functions develop into or determine the other?
- Do the two functions belong to the same functional domain, or is one a constituent of the other?

In accordance with nature, the orgonometric form of the association of *matter* and *energy* has to be either a *paired variation* (4) or some form of *development* (5). Since the number of basic possibilities are few, we can easily try out all the possible arrangements. The following series of experimental equations illustrates most of the possible arrangements:

- If *energy* and *matter* are related as paired variations, we would represent this relationship as follows:

ENERGY ⊥ MATTER 3.04a

Consequently, the whole development equation for the pair would take the following form:

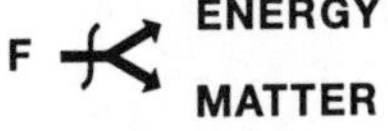

3.04b

- If *energy* and *matter* are not related as a functional pair, but relate as constituents of a development, then the arrangement might appear in any of the following possible ways:

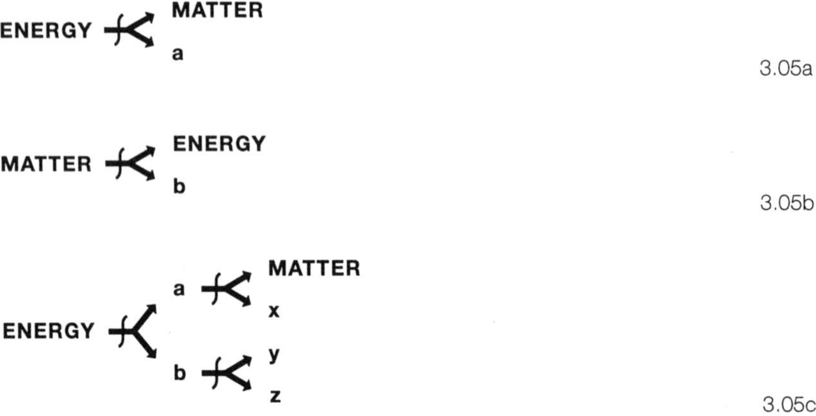

For these experiments, the letter-labels **F**, **a**, **b**, **x**, **y**, **z** represent undetermined functional constituents.

Equations 3.04a and 3.04b are only valid if the given functions are identical with respect to their CFP, that is, if they are directly related as *paired variations*.

Equations 3.05a to 3.05c are only valid if the given functions are indirectly related just as a CFP is related to one of its *variations*. They would then express a natural hierarchy of domains, that is, the functional distinction of a broader realm of operation versus a narrower constituent operation. In this respect, we can immediately rule out experimental equation 3.05b because the arrangement shows **MATTER** to be more primary than **ENERGY**. According to functional logic, this cannot be the case, and the whole formulation is therefore functionally meaningless or "false."

A review of the experimental formulations highlights two closely related issues that will clarify which of the above formulations describes our subject best:

- First, we need to qualify the functional meaning of the given term "energy" in a way that will clarify its precise domain of functioning;
- Second, we need to qualify how we perceived this "energy" in the original associative idea.

These two expressions need to be integrated before we can proceed toward a meaningful conclusion. The *integration of our perception* (subjective response) with *concrete facts* (objective processes) is the hub of the whole technique of orgonometry. The intensity of this integration is determined by the function of *orgonotic contact*. Since this function varies according to the character of each individual researcher, the comprehensiveness of any whole investigative process will also vary. However, as Reich proved, this is the only way we can reach a true understanding of reality (6).

3. The Relationship of Secondary Energy and Matter

If we follow our initial perception, it leads us to think of *matter* and *energy* as a related pair of functions, just as they are associated in classical physics (experimental equations 3.04a and 3.04b). In which case the logic of functions requires that we specify the nature of the term *energy*. Only functions belonging to the same realm of function can be correctly paired. *Matter* belongs to the range of functions we call the *secondary* realm, whereas the function *energy* specifically means the *primordial energy*, that is, a *primary* realm function.

Therefore, the function "energy" in the given idea must really mean *secondary energy*, and this correction resolves the functional meaning of our initial perception. We can now describe the given relationship as follows:

SECONDARY ENERGY ⨍ **MATTER** 3.06

MATTER and **SECONDARY ENERGY** do seem to be a functional pair. They are antithetical, heterogeneous, and yet they also have something in common. Though we have yet to validate this equation, we have already arrived at the *whole functional content* of Einstein's equation via the logic of functions (see equation 3.03). Classical science regards this pair of functions as the "basic duality at the root of nature."

The most important logical test for verifying a *paired relationship* is to find the specific, deeper function that unifies them, that is, their *common functioning principle* (CFP). Let us assume, in the present case, we do not know the specific function that constitutes the CFP and therefore cannot verify the given paired relationship in this way. Nevertheless, we do know that a CFP exists, and can refer to it as *function* **F** (see equation 3.04b).

4. Transformations of Energy and Matter

Evidence of a *functional transformation* between the two functions is another functional way to verify the association of a *heterogeneous paired relationship*. We can cite a number of natural examples of the functional transformation of *matter* into *secondary energy*: Energetic radiations from the sun, the stars, and the radiations of radium and uranium. These were known in 1905, but we have since had further proof via atomic and hydrogen bomb explosions, and via *matter* to *energy* conversions of nuclear power stations.

We can now revise the operational symbol in equation 3.06 to express the direction of this *functional transformation*:

MATTER $\nrightarrow$ **SECONDARY ENERGY** 3.07

The equation states that **MATTER** and **SECONDARY ENERGY** are related as a functional pair, identical with respect to their unstated CFP. Further, it states that **MATTER** can *transform* into **SECONDARY ENERGY**.

Because the two functions are heterogeneous, the above formulation is not necessarily complete. We must still ascertain if *secondary energy* can transform into *matter*. Does the operation of this association express a one-way or a two-way transformation process?

Before we begin to look for concrete evidence of this transformation, functional logic directs us to think it might exist. Since *primary energy* creates *matter*, we could anticipate its secondary manifestation, *secondary energy*, would likewise express a capacity for transforming into *matter*. However, we must still demonstrate that such a transformation can take place. No system of logic can substitute for direct observation of concrete processes and functions.

There is a great deal of practical evidence to show that *secondary energy* does transform into *matter* but this changeover is never as dramatic as the sudden release of energy (nuclear fission), or the glow of radium. The structuralization of energy is a quiet, slow process not easy to observe in the isolation of an experiment, though its aggregated expression is quite obvious in the life process and in the cosmos.

We have evidence of heat-storing chemical reactions that lock energy into a compound structure. There is the living process of photosynthesis that converts light energy into plant food structures. As a further example, we apply the energy of electricity to move particles from one material to another, thereby altering the structure of both. We do not doubt that in some way, *secondary energy* is converted into *matter*. Thus, equation 3.06 can also be revised to show the *direction* of this *transformation* operation as follows:

SECONDARY ENERGY ↛ MATTER 3.08

The equation states that **SECONDARY ENERGY** and **MATTER** are related as a *functional pair*, identical with respect to their unstated CFP. Further, it states that **SECONDARY ENERGY** can *transform* into **MATTER**.

We now have two equations, 3.07 and 3.08, which taken together, complete the paired operation of the functions *matter* and *secondary energy*. Each *functional transformation* requires an individual equation to represent it correctly (7). Technically, the two equations may be combined in a single linear arrangement as follows:

MATTER ↛ SECONDARY ENERGY ↛ MATTER ↛ (etc.) 3.09

However, in the present context, this form does not yield any additional information.

The complete orgonometric setting, equations 3.07 and 3.08, shows that *matter* and *secondary energy* are a pair of heterogeneous antithetical functions, identified by an unspecified *common functioning principle*. Either function can transform into the other but, viewed in isolation, each appears to function exclusively in one state only. We also observe that *matter* and *secondary energy* belong to the same domain of function within the range of secondary functions.

Summary

Thus, we have demonstrated that the functional relation of *matter* and *secondary energy* can be derived through the technique of orgonometry in a way far less tortuous than the path physics took. Further, the orgonometric conclusion, equations 3.07 and 3.08, incorporates more functional information than the classic formula 3.01 or the derived equation 3.02. Equations 3.07 and 3.08 describe the qualitative relationship of the equated functions, and the *direction* of their functional transformations. They also imply both functions are developed from a deeper, more primary function (unstated, but understood to exist).

Along the way, and quite unexpectedly, our functional analysis has exposed the abstract origin of an old stumbling block to comprehensive thinking in the sciences, namely, the assumption that nature develops from a "fundamental duality" of functions. The function *secondary energy* is mistakenly thought to be a *primary* function, and as a consequence, *matter* is also mistakenly understood to be a *primary* function. Reich's discoveries began to contradict this scientific belief almost half a century ago. Whereas, the association of the two functions is a concrete fact, this association is determined by another, more primary function, that is, their *common functioning principle* or CFP.

The orgonometric description of the development of natural functions begins with a single primordial function, and this fact dissolves the "fundamental duality" assumption. What classical physics considers to be "fundamental," is only basic within the functional constraints (or limit) of secondary realm development.

Addendum

5. The Unstated CFP

We cannot expect to validate the association of a pair of functions, without knowing the nature of their CFP. However, according to the logic of the development of functions, we do know a CFP exists, even though its specific qualities remain unknown. This anticipatory knowledge gives functional thinking a powerful advantage over non-functional thinking. In the present case, we know the relationship of *secondary energy* and *matter* is determined by a more primary function, their CFP.

To avoid the lengthy digression required to explain the operation of the actual CFP, we chose to assume the CFP of *secondary energy* and *matter* was unknown, but it could be represented as *function* **F**. On a temporary basis, we could have described it simply as "primary energy." My investigations indicate the *repeating* operation of the *function of creation* is the specific function that integrates *secondary energy* and *matter*. But the presentation of this subject will have to wait for another occasion.

6. What Reich Noted about this Transformation

"We find in nature . . . paired functions which can be transformed into one another. Thus, for instance, matter can be transformed into energy and energy into matter: **M ⇸ E** and **E ⇸ M**. In such cases, we shall speak of *functional transformations* . . . " (8).

7. Transmutations of Energy and Mass in Nuclear Reactions

Functional transformation describes a particular operation whereby one function transforms into another function. This operation is directional (like a *development*), and it only occurs between functions of different kinds, i.e. *heterogeneous* functions. It does not describe the *extent* or degree of the change but the *quality* of the change. Transformations occur in operations of *development*, *fusion*, and *conversion*, and they may be partial or complete.

In 1919, Rutherford bombarded nitrogen with particles emitted by radium and succeeded in transmuting the element. "*Nitrogen was transformed into an isotope of oxygen . . .*"(9). The reaction is formulated as follows:

$$_{7}N^{14} + {_{2}He^{4}} \longrightarrow {_{8}O^{17}} + {_{1}H^{1}}$$

subscript = atomic number
superscript = mass number

It is clear from this conversion that the *mass* of one element was increased, i.e. N^{14} was transformed into O^{17}. Atomic fusion invariably results in products with a greater mass.

Thus, the transmutation of mass (increase or decrease) can be accomplished with an input of *secondary* energy. At the same time, the reaction releases some *secondary* energy. In most of these experimental conversions, more energy goes into the process than is released. Nuclear explosion is an exception, but even so, the atomic bomb conversion only results in a loss of 1/10th of 1% of the mass of U-235.

The classic "law" of the conservation of *mass* and *energy* is not perceived to be breached by the nuclear reaction. However,

Reich's formula for the *function of creation* does appear to breach this "law" — as does the accumulating property of orgone energy, which is a *negative entropy* expression. The classical physicist would therefore see the *function of creation* as another major problem in Reich's work. By contrast, the transmutation of mass is now an accepted fact, and physicists will not readily disagree with the following formulation (see equation 3.08):

(secondary) ENERGY $\nrightarrow$ **MATTER**

8. Does Secondary Energy Really Transform into Matter?

Despite the evidence cited above, we continue to perceive a deeper functional meaning in the accepted observation that *secondary energy* transforms into *matter*. In Reich's formulation (see quote in section 6 above), he does not qualify the meaning of **E** (energy). We would therefore interpret his letter-label **E** to mean *primary energy* rather than *secondary energy*. In which case, the transformation of **E** $\nrightarrow$ **M** (but not of **M** $\nrightarrow$ **E**) describes a deeper functional process, since it represents a transformation across domains of function, from the deeper, *primary realm* to the narrower, *secondary realm* (but not vice versa). Reich accomplished this form of conversion (**E** $\nrightarrow$ **secondary E**) in his orgone energy motor demonstration in 1948.

The *function of creation* continually expresses this process in nature but always in one direction only, toward ⟶ the creation of *matter*, *secondary energy*, *living organisms*, and so on. The *function of creation* may therefore continue to operate in what appears to be a transformation of *secondary energy* into *matter*, i.e. in nuclear fusion. The new matter that emerges from a nuclear reaction may itself express the orgone-*physical* process of the *function of creation* — a primary realm operation — and not a transformation of *secondary energy*.

If this occurs, then the "law" of the conservation of *mass* and *energy* would be breached by such a reaction — more would emerge within the system of the reaction than existed before. In which case, the increased output (energy, mass, or both) might be large enough to be measured.

4

FUNCTION VIEWED AS A CFP

Sections

We must comprehend first the common functioning principle in nature itself if we ever intend to get out of the sterility and great erring.

WILHELM REICH (1)

Research directed toward the primary realms inevitably generates mistakes. But some of these can now be prevented, even before we have uncovered all the constituents of the functional process being investigated. The comprehensiveness of the orgonometric technique makes this possible by providing us with a universal framework for all functional operations, known and unknown.

When an erroneous constituent function is introduced into the setting of a development equation, it alters the entire meaning of the functional process described by the equation. Though the specific error may be quite subtle, it becomes "magnified" in the relation with the other constituents of a *whole* development equation, and this greatly improves our capacity for detecting such errors (2).

Despite its immediate simplicity, the concept of the *common functioning principle* (CFP) is a frequent source of subtle error. This concept is limited by its "reverse" development, a generic characteristic, which differentiates its own functioning from the many objective functions it describes. It simply does not function as a CFP, except in one singular case which we will discuss.

Since basic research always looks toward the left ⟵ of an orgonometric equation, this "reversed" viewpoint continues to be a serious problem for the researcher. However, once this directional characteristic is understood, some errors of thought can be avoided in future orgonometric formulations.

1. Direction in a Concept

Functions and functional processes continuously *develop*, and by agreement, we represent this continuous primary expression as a direction ⟶ toward the right (3).

When a *whole* development is abstracted, as in the form of a *development equation*, two directions emerge:

- A direction ⟶ toward the *complex*, which coincides with development;
- A "reverse" direction ⟵ toward the *simple*, which expresses *the structure of a past development.*

These antithetical directions (or views) express the abstract *operation of thought*, but in this characteristic, the operation of thought is no more than *partly right* since nature develops in only one direction.[1] This technical difference between *thought operations* and *functional development* is the intrinsic reason we must constantly compare the direction of our abstractions with the actual direction of *development.*

Abstract *concepts* that structuralize the process they describe (⟵), such as the concept of the *common functioning principle*, are often used inappropriately. Because they express a "reverse" direction, they cannot be substituted for physical functions in a regular *development equation* or statement. Their characteristic direction contradicts the basic direction of the *development equation* or statement, and this contradiction has to be integrated in some way.

[1] *Partly right* has a precise functional meaning, which will be elaborated in chapter five.

2. The One Direction of a CFP

The concept of the *common functioning principle*, or CFP describes the unifying role of a deeper function from the viewpoint of two known functions that appear to be qualitatively paired. Wilhelm Reich developed this concept after observing a consistent coupling of antithetical functions in nature. Thus, the *concept* and its *name* are strictly derived from an investigation *directed toward the CFP* (⟵), and both continue to express this "reverse" direction.

The functional significance of this "detail" manifests itself when it is abused. For example, the following description of a function *in the role of a CFP*, is functionally *false*:

The CFP determines functions A1 and A2.

In this sentence, the inherent direction of the term "CFP" contradicts its role as the grammatical *subject*. The sentence develops normally ⟶ toward the right, but its *subject* points away from it ⟵ toward the left. The structure of the *whole* sentence is thereby distorted in relation to its subject matter.

The true functional meaning of this sentence, as given, is quite unlike its intended meaning. What it actually says is: An abstract *concept* determines **functions A1 and A2**. But this can only be true if the variations, **A1 and A2**, also represent abstract *concepts*.

To achieve the intended meaning, we can correct the above sentence by subordinating the term "CFP" to a specified function, say *function* **A**, in the role of the grammatical *subject* as follows:

Function A, the CFP, determines functions A1 and A2.

Because of its inherent direction, the term "CFP" functions correctly in the role of the grammatical *object*. Therefore, the first sentence example can also be rectified by reversing its *subject* and *object* as follows:

Functions A1 and A2 are determined by their CFP.

In this sentence, the functional content (i.e. the development being described) is entirely reversed, thereby integrating the inherent "reverse" direction of the term "CFP" with its *role* in the sentence (object). As a functional statement, this sentence *structuralizes a past development.*

Reich consistently observed this characteristic of the *concept* and the *term* "CFP" in his descriptions of its meaning. The "reverse" direction ⟵ toward the CFP is expressed in the awkward structure of a sentence that defines the paired relationship of functions:

Paired functions are identical with respect to their CFP.

Because this sentence describes a "reverse" development ⟵, its structure is functionally limited by the operation it describes. Aside from a few word substitutions, no other sentence structure (in English) will express the functional content of the above sentence.

3. Defining a Concept with an Equation

The sharpness of orgonometric formulations is an exceptional feature of the whole technique. The diagram-like form of a *development equation* significantly simplifies the presentation of this functional process (4). It immediately conveys the *simultaneity* of the interaction and the *simultaneity* of the extension of its constituent functions. It also clarifies the levels of functioning within the *whole* development, that is, the depth of the *realms* and the order of the *domains*. No previous system of thought has ever been so closely aligned with the natural operation of functions and functional processes.

Because it is completely functional, the technique of orgonometry can be used to investigate or describe itself. For example, we can describe the *development* of the *concept of the CFP* as follows:

VARIATION 1 (of 2)
VARIATION 2 (of 2) ≻↦ **COMMON FUNCTIONING PRINCIPLE** 4.01

The equation states that the *perception* of an association of functions (**VARIATION 1** and **VARIATION 2**) is expressed by the unifying *concept* of the **COMMON FUNCTIONING PRINCIPLE**.

This semi-verbal *orgonometric definition* of the CFP correctly represents the development of Reich's concept, that is, the *development of an investigation that led to the concept of a common functioning principle.*[2]

Reich began with the observation that functions are frequently paired in nature. The postulate of a third function as the common source of the pair of functions followed spontaneously. The functional term "common functioning principle" expresses the direction of its conceptual development as represented by the above equation 4.01.

4. The Role of the CFP

In a basic *development equation*, the function we identify as the CFP appears on the left because it precedes the paired functions shown on the right. The development expressed by

[2] Orgonometric word-label equations represent a new technique developed by the author for defining concepts with complete precision.

equation 4.01 has a "reverse" arrangement. However, we cannot simply reconstitute the above formula to match the basic form of a development of concrete functions. If we do this, the entire meaning of the equation alters according to the function of its CFP, which in this case is a *concept*.

Nevertheless, we will temporarily reconstitute equation 4.01 in the form of a regular *development equation* as follows:

COMMON FUNCTIONING PRINCIPLE ⤙< **VARIATION 1 (of 2)** / **VARIATION 2 (of 2)** 4.02

This equation states that the *concept* of a **COMMON FUNCTIONING PRINCIPLE** determines (or develops into) two different *concepts*, represented as **VARIATION 1** and **VARIATION 2**.

Because it is a function that expresses a "reverse" direction, the *concept* of a CFP can only determine other *concepts*. Thus, the functional meaning of the above equation is quite unlike the intended meaning as defined by equation 4.01.

Contrary to our original intent, the developmental facts expressed by the above equation are historically untrue and logically muddled. We know the term "CFP" was derived from a prior observation of an association of variations. But even if we were unaware of this historic fact, the contradictory direction expressed by the term "CFP" itself (⟵) makes the *whole* equation functionally "false" in relation to the agreed *direction of development*. Quite simply, *no objective function can work against the direction of development.*

5. Discovering an Unknown CFP

Discovering an unknown CFP is not like finding a reference in a library. There is no way to *deduce* its unknown qualities from the known characteristics of a pair of functions. We can only *discover* the unknown function through some spontaneous act of research.

Two practical barriers have to be overcome. First, the CFP lies in a deeper, broader realm of functioning than the realm of the known pair of functions — the barrier of the realms. Second, the direction of the research contradicts the natural direction of the process under investigation — the barrier of opposite directions. These barriers are not easily surmounted, even within the functional framework provided by orgonometry.

We need to be aware, in the search for an unknown CFP, that nothing can be taken for granted. Each proposed answer has to be thoroughly tested in its role as the CFP. We cannot be satisfied until the *whole* development is completely *integrated* according to equation 2.15 and the logic of natural functioning (5).

6. Substituting One Concept for Another

One particular mistake frequently appears in word-labeled formulations of a *development equation.* This is expressed in the functional content of the label that purports to be the CFP of a *whole* development. When this error is made, the content of the label translates into "function unknown," or worse, a conceptual "category" masquerading as a concrete function.

Consider the following example:[3]

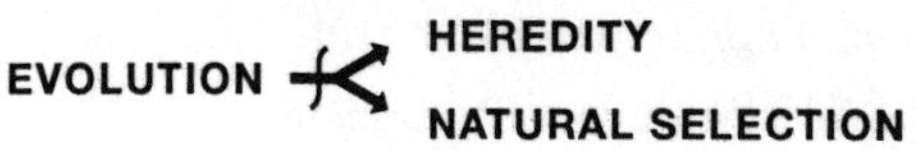

4.03

[3] This proposed equation was presented in a lecture given at New York University, May, 1982.

The equation was meant to show that the paired functions, **HEREDITY** and **NATURAL SELECTION**, are identical with respect to the "function" of **EVOLUTION**.

On close inspection, we find that the term "evolution" expresses a contradictory direction ⟵ toward the left, as for the term "CFP" previously. This characteristic alters the entire meaning of the above equation, and thereby shows that the arrangement is functionally *false* with respect to the intended meaning.

Having formulated this equation, we must then ask:

Does *evolution* determine *heredity* or *natural selection*?
Not clear at this time.
If we change the order of the constituents, does *heredity* determine *evolution*?
Yes.
Again, does *natural selection* determine *evolution*?
Yes.

According to the above answers, the real *direction of development* appears to have been somehow "reversed" in the given arrangement. We can reconstitute it correctly (in this case) as follows:

HEREDITY
NATURAL SELECTION ≻⟶ **EVOLUTION** 4.04

This equation states that the *perception* of **HEREDITY** and the *perception* of **NATURAL SELECTION** are fused in the *concept* of **EVOLUTION**.

We can now see that the term "evolution" functions as a "reverse" *concept*, exactly like the concept of the CFP previously. It cannot be substituted for an *actual* function in a normally directed formulation of a *development equation*.

In the search for a CFP, we must consider the integrity of the *whole* formulation. A "reverse" concept can only function as a *concept*. It cannot substitute for a *concrete function in the role of a CFP* within the setting of a *development equation*. In a regular *development equation*, when the *function* in the role of the CFP is unknown, we are required to label it as "unknown function," "unknown **F**," or simply "?" If we break this stricture, we run the risk of deluding ourselves about the process under scrutiny.

7. An "Incomplete" Function as a CFP

In the process of arriving at a complete orgonometric equation, we may reach a stage where the formulation is complete except for the CFP.[4] Reich presented such an example in "Orgonomic Functionalism" (6). We will briefly review his equation to demonstrate another variation of an unresolved flaw in a stated CFP.

Reich's formula can be reconstructed with word-labels as follows:

$$\textbf{FORCE} \times \textbf{DISTANCE} \prec \begin{matrix}\textbf{MECHANICAL ENERGY}\\ \textbf{WORK}\end{matrix} \qquad 4.05$$

The equation states that the functional pair, **MECHANICAL ENERGY** and **WORK**, is identical with respect to the function represented by **FORCE** X **DISTANCE**.

Again we must ask:

Does *force* X *distance* determine *mechanical energy* or *work*? Yes.

[4] Qualitative formulations are said to be "complete," according to the requirements of equation 2.22 in section 34 of chapter two.

Will any other arrangement of the same constituents work?
No.
But we also note, according to the form of the equation, that the function described by *force* X *distance* belongs to a deeper realm of functioning than either of the two, paired variations.

Reich substituted the standard physical formulas for the constituent functions in the above equation to arrive at the following quantitative orgonometric equation:

$$mlt^{-2} \times l \prec \begin{matrix} ml^2t^{-2} \\ ml^2t^{-2} \end{matrix}$$ 4.06

mlt^{-2} = Dyne
ml^2t^{-2} = Erg

Qualitatively, this equation is exactly the same as the previous word-labeled equation 4.05. But, by substituting the "mathematical expressions" of the constituent functions, we are *simultaneously* including the *quantitative* expression of the *whole* development. An orgonometric equation becomes *complete* when both its *qualities* and its *quantities* are functionally integrated.

As given, the formula representing the CFP is mathematically identical to the formulas of the variations. This clearly demonstrates that the *formula* in the role of the CFP does not express a deeper realm of functioning than its paired variations. It is therefore "incomplete" as a functional statement and must be structuralized further in the direction ⟵ of the primary CFP.

We may note here that it is the *dimension of mass*, represented by **m** in the formula of *force*, which constitutes the functional error or "incompleteness." *Mass* is a mechanical function belonging to the same *domain* as the paired variations.

Summary

The *concept* of the *common functioning principle* can be easily misrepresented when its operating direction ($\longleftarrow$) is neglected.

Concepts that function in the role of the CFP can only determine other *concepts*.

A generalized *concept* that derives from the observation of a partly known process should not be substituted for a *real* function in the role of the CFP. Even for a short time, this "purely" intellectual practice is technically confusing.

The technical evolution of a *complete orgonometric equation* includes stages that are "incomplete" or "pending." Errors of this kind exist for a short time and are an unavoidable consequence of research directed toward the CFP.

Research in the direction of the CFP *structuralizes the past*, an operating quality that frequently contradicts the *development* we *perceive* in the *process under investigation*. This contradiction (an expression of abstract thought) has to be rectified before we can arrive at a comprehensive view of the *process under investigation*.

5

THE FORMULATION OF *RIGHT* AND *PARTLY RIGHT*

Sections

The many-sidedness of my sympathies later led me to the principle that "everyone is right in some way"; it is only a matter of finding out in what way.

WILHELM REICH (1)

The subject of this chapter is an ordinary everyday expression of thought, namely, the antithesis represented by the paired concepts — *right* and *wrong*. We use these concepts without realizing they can and do lead us repeatedly into errors of thought. The pairing of *right* and *wrong* appears to be logical and correct. Nevertheless, we can demonstrate that this conceptual antithesis is functionally flawed. Once these terms or concepts are arranged as an orgonometric equation, their functional characteristics become magnified. We come to understand exactly how their functions are dissociated, and exactly how we can reconstruct a correct and complete representation of the antithesis.

The pairing of *right* and *wrong* is a prime example of the old ways of thinking, which we transmit and perpetuate through the structure of our language. Language and concepts are functionally intertwined, that is, simultaneous. When we separate the two components for any purpose, we immediately destroy their simultaneity. Orgonometric equations are exceptional in that they directly express the simultaneity of functions. This practical expression is exactly what is required for the complete investigation of concepts. It allows for the separate representation of constituent functions within the presentation of a simultaneous *whole* function.

In the process of this investigation, we gradually evolved an orgonometric method for judging the validity of concepts and, simultaneously, a technique for regulating word relationships. These procedures open a path for a deeper comprehension of concepts and language and, in this sense, this chapter can serve as an example for future studies.

1. The Anomaly of Concepts

Every concept, however strange or improbable it may seem, has some functional content. This functional content describes the real "substance" of any concept. In the setting of an orgonometric equation, this functional content emerges as the only active expression or functional meaning contained in any concept.

Despite having a functional content, few concepts are found to be valid when they are tested within the framework of orgonometry. This anomaly continues to be a barrier to deeper comprehension. It expresses the functional difference between the *narrower* operation of "pure" thought and the *broader* operation of concrete functions.[1] This difference is not primarily determined by distorted perception, but is functionally inherent in the operation of pure thought. After correcting any errors of distorted perception, we are still left with errors determined by the functional limitation of the operations of pure thought.

The technique of orgonometry overcomes this formidable difficulty by comprehending the whole framework of natural function — everything is determined by the primordial energy — and by recognizing that we can only comprehend this completely through the function of *orgonotic contact*. To be

[1] Many concepts represent the operation of "pure" thought as distinguished from percepts that originate in organ sensations. Since the isolated operations of pure thought are not always aligned with the operations of nature, we understand such concepts to be generally *partly right* (qv).

valid, a concept must conform to the functional conditions set out in equations 2.14 and 2.15 in "Basic Orgonometry" (2). These equations tell us a valid function (or *function in a concept*) will always express the complete integration of the *thought process* and the objective process. The function *nature*, their *common functioning principle* (CFP), determines what this means in its entirety.

2. Observations about Method

In practice, we need to validate the functional content of each concept simultaneously as a development and as a functional constituent of a pair of variations. This procedure cannot be formalized, as in a mathematical proof, since it will vary according to orgonotic contact. Each of the two constituent functions, the process being investigated and the process of the investigation, are determined by the deeper function, orgonotic contact. We may assemble some general rules for all orgonometric investigations, but each investigation must follow its own formal development (3).

The investigation of a given concept generally begins with the uncovering of the function in the concept. We then try to find its true functional variation or antithesis. Until this is accomplished, we cannot proceed to develop an integrated picture of the *whole* function as expressed by the paired concepts.

In practice, we find we do not need to refer to the CFP of paired concepts except as an abstraction. This appears to be a consistent technical fact shared by all paired concepts. We cannot elaborate this observation here, but it does explain why concepts can be completely investigated and resolved in the form of paired variation equations.

At the beginning of such an investigation, we usually do not know whether the given concept is valid or false. Practically, it is better to assume it is flawed in some way rather than to assume the opposite.

3. *Right* Paired with *Wrong*

Since we use the words *right* and *wrong* as an antithetical pair of concepts, they can be arranged in the form of a basic orgonometric equation for paired functions as follows (4):

RIGHT ⨍ **WRONG** 5.01

The equation states that the concept **RIGHT** and the concept **WRONG** constitute a pair of functional variations. It also implies that **RIGHT** and **WRONG** are identical with respect to an unstated CFP.

The specific operational symbol that links the two concepts indicates the paired relationship is either very *simple*, or as yet undetermined, and this leads us to the next step.

4. Functional Content

The terms *right* and *wrong* represent concepts with a functional content. However, when placed in the context of a paired relationship, some specific characteristic of the constituent words (or concepts) may falsify the functional integrity of the association. We need to investigate such an operation as a pair of functions before we can arrive at a true picture of the concrete relationship represented by the words (or concepts).

According to their everyday use, *right* and *wrong* function as antithetical opposites. For example, 2+3=5 is described as exclusively *right*, and 2+3=6 is described as exclusively *wrong*. Further, *right* and *wrong* are expressed as if they were functionally *homogeneous*, that is, the same kind of function but of extreme opposite meaning.

Thus, in their everyday use, *right* and *wrong* operate as a mutually exclusive, antagonistic pair of opposite variations. Yet the dissociated etymology of the word *right* and the word *wrong* subtly contradicts this practical association and alters its meaning. Before we can formulate the paired functions correctly, we must somehow dissolve this contradiction.

5. Abstracting the Paired Relationship

We can abstract the operation that describes the relationship of the concepts of *right* and *wrong*. This operation is represented by a basic orgonometric equation for antagonistic opposite functions as follows:

$$\mathbf{A1} \nleftrightarrow \mathbf{A2}$$ 5.02

The equation states that function **A1** and function **A2** are identical with respect to an unstated CFP. It also shows that the paired functions express antagonistic, mutually exclusive, opposite characteristics.

The equation entirely agrees with the way we apply the terms *right* and *wrong*. If **A1** represents *right* then **A2** represents *wrong*. We also note the same letter-label **A** appears on both sides of the abstract equation, and this indicates the homogeneous quality of the two functions. Only homogeneous functions can interact in this specific way.

6. A Functionally False Formulation

We may now try to restate equation 5.01, so that it will conform with the preceding abstraction:

RIGHT $\nleftrightarrow$ **WRONG** 5.03

This equation appears to say that **RIGHT** and **WRONG** are functionally identical with respect to an unstated CFP, and that they operate as an antagonistic, mutually exclusive pair of homogeneous functions.

If we consider the formulation more carefully, we can observe that the terms, *right* and *wrong*, are not at all homogeneous. They are not of the same kind, according to their derivations or their roots. Within the system of language, they do not accurately represent the homogeneous characteristic expressed by **A1** and **A2** in the abstract equation 5.02.

In orgonometric equations, precise functional descriptions are crucial for the development of more complete statements. Just as a complete orgonometric equation requires both function and number to be correctly integrated, so too do the word-labeled forms of the equation. Before we can add the specific operational symbols that define a relationship such as the one we are considering here, the quality of the terms we use must also correctly reflect the basic functional relationship.[2] The above equation is therefore technically *false,* because we have used a pair of *heterogeneous* terms in a *homogeneous* arrangement.

7. The Functional Inconsistency of Words

We may find that an intuitive appreciation of functions and functioning regulated the original structure of the earliest forms of language. But with the establishment of man's

[2] The words express concepts, and distorted concepts become entrenched in the way we use words.

chronic condition, *armor*, the development of language altered as a *whole*, and its structure began to express the functioning of man's armor. We should not be surprised to discover a large proportion of the words we use are not functionally integrated, according to their associative content and to their objective origins.

The dissociated character of individual words is only one of the practical problems to be considered when we investigate concepts with the technique of orgonometry. Another is the built-in "reverse" characteristic common to structuralized concepts about the past (5). Since words and concepts are inseparably linked, the gradual development of orgonometry will force us into reconsidering many established word pairs, including standardized antonyms and synonyms.

8. Correcting the Distortion

The etymology of the word *wrong* is quite different and separate from that of the word *right*. They do not share a common root, which would define them as being directly related and homogeneous. For example, the pair *correct* and *incorrect* expresses an obvious homogeneous functional relationship, whereas the pair *right* and *wrong* does not.

Others, like *freedom* and *law* or *true* and *false*, are directly related by their agreed meaning, that is, as concepts. But they are not directly related as concrete word functions, that is, as units of language.

The study of the functional content of *concepts* simultaneously requires that we examine the *words* we use. The intrinsic properties of the words themselves often distinguish those concepts that are the creations of pure thought from those based upon reality.

9. Correcting the Word-equation

By altering one of the words in equation 5.03, we can correct that statement and make the words and the functional content completely homogeneous. At first, it seems we have two alternatives: One in which *right* is retained; and another in which *wrong* is retained. However, as we shall see, this dual option expresses the characteristics of a pure thought operation rather than the logic of concrete functions.

Let us consider both options for a while, opposing each prime term with a negative qualification as follows:

RIGHT $\not\leftrightarrow$ **NOT RIGHT** 5.04a

WRONG $\not\leftrightarrow$ **NOT WRONG** 5.04b

Each of these equations correctly describes the homogeneous characteristics of two antagonistic opposite conceptual functions. Both represent the functional content of the subject of this investigation more accurately than the previous word-labeled formulations (equations 5.01 and 5.03).

We can now observe how the two formulations differ. The first expresses a positive view (characteristic of optimism), while the second expresses a negative view (characteristic of pessimism). Since functional facts are not decided by personal preferences, we must try to find a concrete functional reason for determining which one, if any, is the correct formulation.

10. Determining Priority among Concepts

In the above two equations, and at this stage of our investigation, we are satisfied that the characteristics of the paired terms (i.e. word-labels) correctly represent the operation of the subject being investigated. However, we must now query the existence of two separate answers. There can only be one completely correct functional answer in which the terminology and the functional meaning are fully integrated.[3] We therefore understand that at least one of the above arrangements must be flawed in some way. Thus, the two equations above show we must still resolve a basic question about the functional meaning of *right* compared with the functional meaning of *wrong*. Which of these two concepts is functionally complete, and which is functionally meaningless or false? Which one of the concepts is definite and which is not? In the earliest exploration of this subject, my reasoning followed this path:

- Judging by their everyday meaning, *right* and *wrong* are often expressed as *absolutes* either simultaneously or at different times. Yet, such an extreme pair of conditions cannot be found in nature or in orgonometric equations. Two *absolute* antagonistic opposites can only be an expression of pure thought. The formulation in an old riddle illustrates the dilemma of such pure, extreme, and mutually excluding concepts:

What happens when an irresistible force meets an immovable object?

According to functional logic, the function of an "irresistible force" would preclude the existence of a function that might be described as an "immovable object." Since everything moves, the concept of an immovable object is functionally meaningless anyway. Given that only one of a pair of antagonistic concepts can be an absolute, then its opposite

[3] In some instances, our language has developed more than one pair of words to represent the same functional association. True alternatives will not alter the basic functional content (or meaning) of one correct formulation.

must be a relative concept. Since *right* and *wrong* cannot both be absolutes, one of them must be a relative concept. If *right* is the absolute, then *wrong* is the relative concept, or vice versa. The *right* answer to any single question is a single entity, whereas a *wrong* answer is one of a multitude of entities. *Right* expresses a characteristic that can be associated with an absolute condition, whereas *wrong* expresses the characteristic of a range, that is, a relative condition.

A second and later argument followed this path:

- We cannot know that anything is definitely wrong unless we know the right answer. Without knowledge of what is right, *wrong* becomes an *indefinite* quality. Therefore, the concept of *wrong* is entirely dependent upon the concept of *right*. Further, when we know the right answer, it can be defined as a single entity, just like the primordial function. Wrong answers take a thousand forms, an endless variety, and only have meaning in relation to how close they come to the right answer, or how they differ from the right answer.

In both arguments, the function *right* emerges as the defining function or concept, while the function represented by *wrong* is both indefinite and derived. Therefore, of the two formulations above, only equation 5.04a, with the paired terms *right* and *not right*, correctly represents the functional content of the subject under investigation.

11. Adjusting the Conclusion

By standards of pure logic, the above arguments may seem to be far from airtight. However, they are quite sensible and must be functional, since they led toward the complete functional solution.

We can now review equation 5.04a again to see if there are any subtle alterations that might be made:

RIGHT $\nleftrightarrow$ **NOT RIGHT** 5.04a

We note that **NOT RIGHT** is a negative expression, and we wonder if there is another way of qualifying the word *right* in a positive way without altering the functional expression we have so far determined. *Partly right* seems a good possibility. We restate the equation:

RIGHT $\nleftrightarrow$ **PARTLY RIGHT** 5.05

RIGHT and **PARTLY RIGHT** continue to describe a pair of antagonistic opposite functional concepts, and the terms themselves are also related in the same way. Both the terms and the functional content have been completely integrated.

The quality in the term *right* features in each of the paired word-labels, just as the quality in the letter **A** features in each of the paired abstractions **A1** and **A2** (equation 5.02). Compare this with other functionally valid word pairs such as *finity* and *infinity*, or *correct* and *incorrect*. The same root word (i.e. quality) also features in each of these paired constituents.

12. First Confirmations

We can immediately test the practical accuracy of this formulation by applying it to a simple concrete example, such as the one used previously:

Does *right* describe 2+3=5 and does *partly right* describe 2+3=4?

Yes. 2+3=5 is entirely *right*.

Can we say that 2+3=4 is *partly right*?

Yes. 4 is part of 5, and actually very close to 5.

What of 2+3=1?

Yes. 1 is also part of 5 and therefore part of being right.

What of 2+3=105?

This too is *partly right*.

Everything "wrong" is functionally *partly right*. It must be so, because *everything that exists has some functional content*.

13. The Case of 2+3=0

Can we demonstrate why we should consider 2+3=0 to be *partly right*? We begin with the knowledge that the concept *zero* is an abstraction that cannot be found in a functional setting (6). However, when units or numbers represent real things, *zero* assumes a limited functional meaning, namely, the *absence of such units* or *none of such things*. Units that represent things do not necessarily represent one unique thing, or even one kind of thing. By qualifying each of the units, we can reconstitute the given equation in a functional and meaningful way as follows:

2 APPLES + 3 ORANGES = 0 BANANAS

This formulation is entirely *right* within the limitations of the logic of sets. No **BANANAS** are present in the combined sets of **APPLES** and **ORANGES**. The fact that it is correct in some way proves the basic formulation (2+3=0) is not entirely "wrong." It must, therefore, be at least *partly right*.

To interrupt briefly, we note that what is functionally incorrect about the above formulation does not concern the mixing of units, but the exaggerated expression of an extreme isolation of the constituent functions within the operation of the formula.[4] Mixtures of different kinds of units are to be found in almost all practical formulas such as $E = mc^2$, ml^2t^{-2}, mv, and so on.

14. The Case of a Spermatazoa Sample

A sample of live, human spermatazoa contains both *right* and *partly right* cell variations. All are living, moving, and therefore functioning. However, since we begin with a general idea of their *whole* biological function, we can distinguish some cells that are definitely *partly right* from those that are *right.*

We observe the motility of the cells first, because this quality impresses us immediately as we look down the microscope, or at the screen. But not all of the cells are moving with the same vigor or determination. One cell may be propelled in continuous circles, unable to target its motion. This cell is clearly *partly right,* and we may further observe it has two functioning flagella. We observe other cells that are sluggish, or irregular in their movements, and they too are physically deformed in various ways. Again, these cells are *partly right.*

The expressed vigor and the form of entirely *right* spermatazoa are quite unmistakable. Nevertheless, in terms of their destination and fulfillment (continued development), only one *right* sperm cell will fertilize one egg cell.

15. The Case of the Extinction of the Dinosaurs

Dinosaurs roamed the earth for 100 million years. As constituent functions of the *whole,* they were *right* for all that timespan. Then the *whole* altered suddenly, perhaps because

[4] According to their biological function, *bananas* are related to *apples* and/or *oranges,* which makes the above statement *partly right* with respect to this association.

a massive meteorite hit the earth, and the constituent function *dinosaur* was transformed in relation to the *whole*, from being *right* to being *partly right*.

We have to admit that during the period of their decline (estimated to be about 200 years), the dinosaurs continued to function as individual organisms, moving, eating, and reproducing as before, within the limitations of their altered environment (the *whole*). So they were never entirely "wrong," even as they were becoming extinct.

Comprehending the functional association *right* and *partly right* becomes very important for research concerned with *past, structuralized development*, as in the study of evolution or natural history. The descriptive label, *partly right*, realigns our "pure" thought processes every time we apply it.

16. A Review of Our Technique

The investigation described in this chapter was entirely built upon functional information conveyed by paired variations placed in an orgonometric equation. We were never forced to explore the precise qualities of the CFP of our paired subject. Because of this, our procedure becomes a demonstration of the depth of functional meaning expressed in a paired variation equation — the simplest form of an orgonometric equation.

The abstract operations of all paired functions were fully described by Reich and reiterated in "Basic Orgonometry" (4). Thus, any given pair of functions must conform to one of these abstract equations. If a given pair of functions does not appear to conform, then we should assume the given statement is *partly right* and therefore flawed in some way. Technically, we would look for the error in: The association; the individual constituent functions; or the operation as a *whole*.

Summary

By applying the technique of orgonometry, we have succeeded in demonstrating that the commonly expressed relationship of *right* and *wrong* is functionally false. We have also exposed the immediate source of the error, which is the expression of the dissociated word "wrong." This was then rectified by constructing a new term, *partly right*, to replace the word "wrong." Accordingly, the functional content of the given antithesis was reformulated as the antithesis *right* and *partly right*.

To test the validity of this conclusion, we applied it to a set of simple arithmetical statements, to immediate observations of individual variations among spermatazoa, and to a past, structuralized event — the interaction between an extinct species, the dinosaurs, and their environment.

Postscript

17. Reich's Conclusion

Some time after arriving at the above conclusion, I associated the concept of *partly right* with an often repeated observation made by Reich:

"Everyone is right in some way." (7, 8)

Partly right accurately represents the essence of Reich's observation. We are once again impressed with the functional precision of Reich's statements. This was no ordinary remark, but a correct observation about an everyday conceptual error.

Reich also made a number of strong statements about the functional characteristics of the concept *right*:

> "There is for one and the same fact only ONE explanation which is objectively correct; there are not ten different correct views." (9)
>
> "There is only ONE correct arrangement and not four or sixteen." (10)

These statements reinforce our observations about the functional characteristics of *right* and our reasoning that it is the root word, as well as the root concept, of the given antithesis. *Right* does not need any further qualification, because it is one condition, whereas *partly right* is an endless range of conditions.

18. "Absolute" and "Relative"— Exaggerated Concepts

We first concluded that *right* was the correct finite term for the finite characteristic expressed in the paired concepts of *right* and *not right*, by associating this characteristic with the conceived expression of an absolute (see section 9). *Right* appears to function like an absolute. *Partly right* (or *not right*) expresses the characteristic of a relative concept, which thereby confirms the practical formulation of the paired relationship.

Later, we came to comprehend that the concepts, *absolute* and *relative*, are essentially exaggerated creations of pure thought. Functions are neither absolute nor relative, and these characteristics cannot be found in natural functioning or orgonometric equations. Like *cause* and *effect*, they only exist as pure thoughts that *derive* from functions or processes (11). Nevertheless, we managed to find the correct solution to our functional investigation by referring to two incorrect (i.e. functionally false) concepts. How then can two incorrect concepts lead us to a correct conclusion?

19. Two *Partly Rights* Can Make a *Right*

Can two *partly rights* make a *right*? The answer begins with the observation that neither concept is entirely "wrong." They have to be *partly right* as we demonstrated previously. *Every concept has some functional content, and this content makes every concept right in some way.* If we abstract the *right* content, the abstraction will be entirely *right*. Thus, the *right* abstraction of two *partly right* concepts can constitute a real base for a correct functional observation.

Conclusion: Two *partly right* expressions can lead to the right answer. Contrast this conclusion with the old saying:

Two wrongs do not make a right.

It now becomes obvious from the functional logic of the above discussion that this old saying is functionally false. However, let us elaborate what we now comprehend. In this familiar saying, we observe the questionable word "wrong" being used in a relation with the word *right*, and this signals a possible functional error. According to our formulation (equation 5.05), the expression "wrong" should be replaced by the expression *partly right*. If we amend the old saying by replacing the word "wrong" with *partly right*, the functional meaning of the old saying becomes much clearer:

Two partly rights do not make a right.

We observe that this reconstruction has dissolved the absolute connotation expressed by the original sentence. However, it also exposes the contactless logic that originally determined the old observation, because two *partly right* expressions must have some *right* content, and it is this content that makes a *right* conclusion possible. Thus, the reconstruction demonstrates the original saying was functionally meaningless or *false*, that is, only *partly right*.

20. A New Saying

We can now create a new statement that is functionally correct and genuinely radical in its expression:

Two partly rights can make a right.

This statement is confirmed by the correct conclusion of the *partly right* arguments of section 10. It can also be verified with a simple arithmetical example as follows:

	2+3 =	**6**	**PARTLY RIGHT**
and	**4+6 =**	**9**	**PARTLY RIGHT**
	total	**15**	**RIGHT**

The whole calculation shows how the sum of two **PARTLY RIGHT** sums can nevertheless provide the **RIGHT** overall total sum.

Orgonometry is primarily concerned with comprehending how everything functions or malfunctions. Our aim is to proceed toward more complete functional statements. Along the way we spontaneously observe errors entrenched in the old ways of thinking, but we do not focus our attention upon them simply to be critical. When we examine these spontaneous observations more closely, they can often lead us toward new, creative formulations.

6
LANGUAGE AND THE DEVELOPMENT OF FUNCTIONS

Sections

1. General Descriptions of the Subject
2. Transposing Development Equations into a Linear Form
3. Comprehensive Qualities Lost in the Transposition
4. The Functional Limit of Language

No, this is not a philosophy, as many of my friends believe. On the contrary, it is a tool for thinking that one must learn to use if one wishes to explore and deal with the living . . .

WILHELM REICH (1)

As a rule, the construction of sentences and equations have a lot in common, but the branching form of a development equation is unlike the monolinear structure of a sentence. Nevertheless, a development equation can be technically rearranged to resemble the structure of a sentence. When this is done, we can compare the original development equation with the rearranged, monolinear version. The comparison reveals, for the very first time, the kind of functional information that is lost in the process of this translation.

What is lost in the technical, orgonometric translation is also lost in the organization of sentences that describe a development of functions. By experimenting with a series of sentences that attempt to describe the functional information expressed in a given development formula, the discussion proceeds to illustrate how such descriptions can be written so they will correctly represent the functional order of the constituents. Simultaneously, this practical analysis also reveals the intrinsic functional limit of all languages.

1. General Descriptions of the Subject

No matter what is being described, the operation of spoken and written language always proceeds in the same way. Words follow words, sentences follow sentences, and paragraphs follow paragraphs. The written form is an arrangement of successive individual words or single constituent functions, and we can represent this sequential operation in a condensed abstract setting:

A1 → Av → Ap → A2 → Ac → A3 → (etc.) 6.01

The letter-labels **A1**, **Av**, **Ap**, **A2**, and so on, represent words. Specific words will be substituted later. The formulation shows a sequence of homogeneous functional variations (i.e. words) that alter from stage to stage. The number of constituents and the number of stages are identical. There is no visible sign of branching.[1]

The pattern of the *development* of functions is quite different from the above. Orgonometry represents it with the following basic equation (2):

A1 ⤙ **A2** / **A3** → **etc.** 6.02

A further stage of *development* would extend the formulation as follows:

A1 ⤙ [**A2** ⤙ **A4** / **A5**] / [**A3** ⤙ **A6** / **A7**] → **etc.** 6.03

Qualitatively, the whole pattern expresses a characteristic of organic "growth," as exemplified in the linear branching of trees and the mitotic reproduction of protozoa. We also ob-

[1] The typographical format (horizontal straight lines) has no special meaning in this context. We could show the formula meandering without altering its basic operational characteristic.

serve, as we look toward the right ⟶ of the equation (the direction of development), that each stage of differentiation doubles the previous number of constituents. After only two stages of development, function **A1** has proliferated into four related but different functions. In a realm where all the developed functions coexist (e.g. in the primary realm or in the expression of structure), the same development would represent a total of seven constituent functions.

The following diagram extends the operation of *development* by a few more stages to emphasize its practical dynamics:

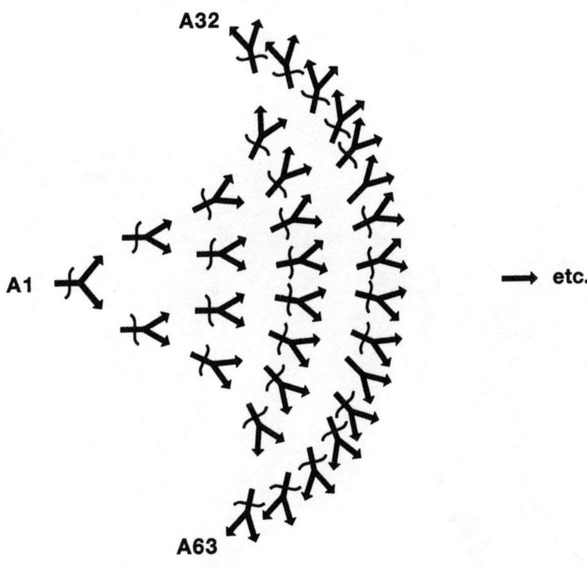

6.04

Qualitatively, we see at once how the simple pattern of development can lead to an intense concentration or an expansive proliferation of simultaneous constituent variations after only a few stages of development. Quantitatively, the products are extensive. For example, starting with one amoeba (the CFP), thirty stages of undisturbed mitotic reproduction would produce 536,870,912 related amoebae. Compare this orgonometric diagram with any one of Reich's earlier functional illustrations of the general process of natural development (3).

If we had to describe the complete development presented in diagram 6.04 with words alone, we would logically choose to describe it in a consistent direction, preferably in the direction of development. Such a chain of words would normally be represented as a monolinear succession of changes, either horizontally or vertically, according to the custom of written languages, or in relation to the chosen time axis in a Cartesian representation. With a poet's license, we could deliberately bend the sequence and make it express the expanding shape of diagram 6.04 in the following diagrammatic way:

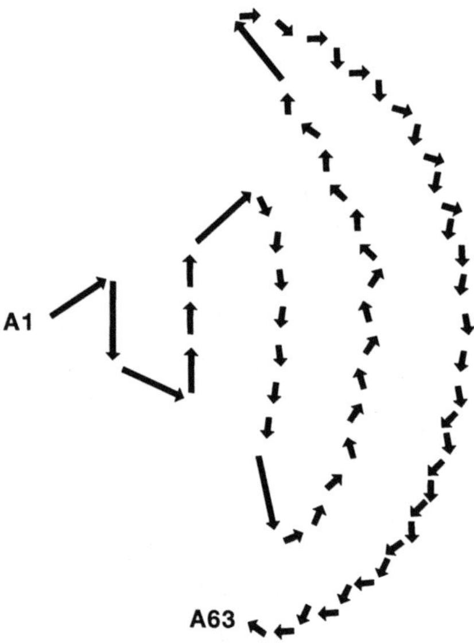

6.05

In the case of 30 mitotic stages of development, we would need something like a billion labels simply to identify all of the constituents, and still more to describe all of the relationships and stages. It would be difficult to arrive at a concept of the *whole* development from such a volume of detail.

When we compare the general picture of a functional development (equation 6.03 and diagram 6.04) to the operation of language (diagrams 6.01 and 6.05), we cannot avoid their marked qualitative difference. The operation of language appears to be monolinear, confined, and constant, whereas the operation of development generates many lines of connection and its simultaneous expression becomes increasingly more complex toward the right. Presented in this way we can clearly see that *the operation of language is limited with respect to the operation of development*. Since the deeper system encompasses the narrower one and not vice versa, we can broadly conclude that the operation of functional development can never be fully described with words alone. Any word description of development would have to be augmented with sensory aids that go beyond the limits of language (e.g. diagrams or gestures).

2. Transposing Development Equations into a Linear Form

Having reached a general understanding of the *whole* of our subject, namely, the operation of language compared with the operation of functional development, we can now begin a more specific investigation of the functional limitations of language by applying the deeper technique of orgonometry.

Two questions emerge: Can we rearrange the form of a development equation into a monolinear form to resemble more closely the general operation of a sentence of language? If so, do we forfeit any qualitative information (i.e. expression) by doing this? The answer to both questions is yes. To demonstrate this, we must first introduce the technical procedure itself, since it has never before been discussed, though we know of two applied examples in Reich's published works (4,5).

The development of a *primary* differentiation is described by the following abstract orgonometric formula due to Reich (6):

$$\mathrm{N} \not\prec \begin{array}{l} \mathrm{Vx} \\ \mathrm{Vy} \end{array} \qquad 6.06$$

This can be rearranged as a monolinear statement:

$$\mathrm{N} \not\rightarrow (\mathrm{Vx} \not- \mathrm{Vy}) \qquad 6.07$$

The basic example of development given previously, equation 6.02, can be formally transposed and restated as follows:

$$\mathrm{A1} \not\rightarrow (\mathrm{A2} \not- \mathrm{A3}) \qquad 6.08$$

Likewise, the extended development equation 6.03, can also be transposed into the following monolinear form:

$$\mathrm{A1} \not\rightarrow \{[\mathrm{A2} \not\rightarrow (\mathrm{A4} \not- \mathrm{A5})] \not- [\mathrm{A3} \not\rightarrow (\mathrm{A6} \not- \mathrm{A7})]\} \qquad 6.09$$

Equations 6.07, 6.08, and 6.09 are monolinear representations of the three given development equations. The braces, brackets, and parentheses have been introduced to retain, as far as possible, the functional hierarchy expressed in the one-becomes-two development process. They define the boundaries of each *domain* and its *realm*—information crucial for a correct understanding of functional development.

We can see from these monolinear transpositions that the operation of a functional development can be superficially represented with a technique limited to a monolinear succession, such as language or classical mathematics. But we note that the order of the domains is not immediately apparent in this setting. We also begin to see how easily its functional definition could be confused by a missing bracket or a "minor" change in the given sequence of its constituent functions.

3. Comprehensive Qualities Lost in the Transposition

My experience with applied orgonometry has led me to conclude that the expanding layout of the development equation (entirely due to Reich's perception of natural development) provides a deep advantage over all other descriptive techniques limited in their mode of operation to a monolinear succession. I find this specific limitation expressed in the way we think, in our languages and in mathematics. And it must also be expressed in any linear transposition of a development equation, such as those we have presented above (equations 6.07, 6.08, and 6.09). By comparing our examples of linear transpositions with their original expanded forms (equations 6.02, 6.03, and 6.06), we can observe those comprehensive qualities that have been lost in the process.

Monolinear equations fail to express a number of important developmental qualities, namely:

- We cannot immediately discern the inherent quality of "growth" with respect to its form, although we can instantly see this in the form of a development equation. Therefore, we need to study the content of a monolinear equation closely before we can know that it represents a functional development;
- We cannot tell which constituent function represents the CFP of the *whole* development. In our examples, the CFP is logically presented as the first function on the left of each monolinear formula but, for the most part, this is seldom the

case. Most formulas (or statements) require close study before we can ascertain which (if any) of the functional constituents is the actual CFP of the *whole* development;

- We cannot see the order of the *domains*, the order of the *realms* (if any), nor the *extent* of the development (i.e. the number of domains or stages of development);
- We cannot discern which *domains* are related to each other without reconstituting the formula as a development equation.

These limitations essentially concern questions about the order of functions — which are broader and which are narrower; and the common domains of those functions that belong to the same domain of functioning — associated and disassociated functional simultaneity.

The operating order of a monolinear sequence is simply too narrow to describe the simultaneities of relationships that coexist in a development of functions. We cannot hope to comprehend a broader process (functional development) from a description that utilizes a narrower technique (monolinear description), and we can now see why the invention of orgonometry was a necessary adjunct to Reich's further explorations of the *primary* functions of nature. He had to devise a descriptive technique that could branch or differentiate, that is, develop, in the same way functions develop in nature. He succeeded and his creation, orgonometry, is broader than any previous descriptive technique.

4. The Functional Limit of Language

All languages operate as monolinear sequences, and all languages are fundamentally limited to this mode of operation. The great subtleties of words and phrases that have evolved in a language, such as English, do not alter this intrinsic functional fact. However, this limitation is common to the whole realm of function that expresses the operation of *"pure" thought*. Thus, the operation of language (an abstract function) is functionally determined and limited according to the operation of thought (a natural function).

We can see from the abstract orgonometric transpositions, equations 6.07, 6.08, and 6.09, that a functional development is incompletely described with a technique that operates without branching. Our analysis (section 3) effectively catalogues all the functional qualities expressed in a development, or an orgonometric representation of a development, which lie beyond the expressive limit of a monolinear technique. Since the operation of all languages is monolinear, we conclude that no language can describe a functional development completely.

When we set out to describe a development equation in words, we are virtually transposing a multilinear operation into a monolinear form. This action is a simple variation of what we did with the abstract representations in section 2. For example, we would describe equation 6.06 with the following sentence:

N differentiates into Vx and Vy. 6.10a

or more fully:

The primary force of nature (N) differentiates into variations (V) that are expressed as pairs of specific functions, which we represent as x and y. 6.10b

The formal similarity between the transposed equation 6.07 and the organization of sentence 6.10a is immediately obvious. In sentence 6.10b, the inclusion of descriptive qualifications, concerning the terms (i.e. functions and functional principles)

and the practical dislocation of **x** and **y** from **V** with respect to word sequence, makes its real similarity with transposed equation 6.07 much less obvious.

Equation 6.02 can be briefly described as:

A1 develops into A2 and A3. 6.11a

or alternatively as:

A1 transforms into A2 and A3. 6.11b

The formal similarity between these sentence variations and the transposed equation 6.08 is obvious. In sentence 6.11b, the orgonometric symbol ⇸ accurately translates as "transforms into" and the symbol + is roughly represented as "and." In sentence 6.11a, the verb "develops" correctly describes the functional symbol ⪪ seen in the original formulation, development equation 6.02. But this branching symbol is excluded from monolinear statements, and the practical substitute, a transformation symbol ⇸, is somewhat ambivalent in the given context (equation 6.08). Technically, the required change of symbol is entirely consistent with a change in comprehensiveness. It also expresses the reduction that follows a transposition from a broader to a narrower domain of functioning (see section 3).

We can now briefly explain the letter-labels used in the first formulation of this paper, arrangement 6.01. The arrangement describes the general monolinear operation of either of the above two sentences (examples 6.11a and b). **Av** represents the verb (either *develops* or *transforms*), **Ap** represents the preposition (*into*), and **Ac** represents the conjunction (*and*). However, these grammatical distinctions are subservient to the broader function expressed by the sequential flow of the word-units in a language.

Summary

We began with a broad observation about language being limited to a monolinear or non-branching operation, a functional limit that becomes especially critical when attempting to describe the *process of development*. Because of its origin, the technique of orgonometry does not have this limitation, and it can fully describe the relationship of functions in a functional development. Therefore, we used the broader technique to study this specific limitation of language.

We introduced, for the first time, the technical procedure for converting the bifurcating form of a development equation into the more commonplace form of a monolinear equation. Then, by analyzing the qualitative consequences, we began to see and describe all the functional qualities this process excludes, because they cannot be conveyed by the narrower technique, i.e. the monolinear form. These losses of comprehension were ascribed to the whole domain of *pure thought* rather than to any one of its various products, such as language or classical mathematics.

We lastly demonstrated the simple, formal similarity between monolinear orgonometric statements and the structure of sentences that describe the same set of functional relationships.

7

ART AS A PROCESS

The Function of Artistic Creation

Sections

1. Art as Creation
2. The Function of Artistic Creation
3. The Significance of Realms and Domains
4. Functional Observations about Artistic Creation
5. Applying the Function of Artistic Creation
6. The Impact of Photography on Modern Painting
7. Kandinski Changes Everything
8. Quotations from Kandinski
9. Paul Klee's Expansive Development
10. The Impact of this New Direction
11. Understanding a Mass-free Orgone Expression

It is not easy to arrive at a conception of a whole which is constructed from parts belonging to different dimensions. And not only nature, but also art, her transformed image, is such a whole.

PAUL KLEE (1)

We cannot avoid noting that the art we know and admire was produced by *armored* persons living and working within an *armored* society.[1] Of all the great artists of the past, I can only point to one or two whose personal history suggests they were blessed with a fair degree of health in the way orgonomy defines health. In this rare group, I would venture to name only one example, William Blake, whose strangeness continues to puzzle even those of us who most admire him. The reader may think of other exceptions, but generally, we have to accept the fact that the celebrated artists of our culture were armored characters who frequently expressed themselves in very neurotic ways.

Nevertheless, these armored individuals managed to tap the deepest resources within themselves and express their deep insights through their art in a way that stirs the same feelings within ourselves. How then were they able to bypass the limitations of their armor without distorting this deeper material? Wilhelm Reich postulated the existence of "holes" in the armor through which most of us can make some contact with our functional core.

But a further question still remains: Why is it some artists, like Shakespeare, Bach, Blake or Klee, were able to elicit a

[1] Some prior orgonomic knowledge is assumed. The term "armor," for example, describes the *whole* function of a biopathic process uniquely associated with mankind (2, 3).

steady stream of deep insights seemingly on demand? We have no reason to suppose they were significantly less armored than other people. Yet, according to their output, they wasted little time waiting for moments of inspiration.

I will try to show it is the operation of the *function of creation* within the work process, which spontaneously and repeatedly exposes the artist to the deeper realms of function. And the artistic work process is the central theme of this chapter.

I have known about the functional priority of *process* in the method of art for more than three decades. But I became convinced there was no specific function for art other than the general function of *work*. Instead, I distinguished artistic work according to its *purpose*, and the purpose, as I understood it, was *to imitate the natural function of creation*. This approach served me well for many creative years.

Then I came in contact with Wilhelm Reich's later writings on *orgonomic functionalism* and *orgonometry*. I must have been ready, because the content poured into me as I studied the material. Toward the end of this exciting period, I spontaneously associated what I already knew about "creation in art" with Reich's formulation for the *function of creation* in nature. The preliminary outline of the "function of artistic creation" emerged a few moments later. However, it took me many months to accustom myself to the objective fact that art does have an inherent function.

Thus, although the *function of artistic creation* can now be described in words, it was first derived as an application of orgonometry, and I doubt if it could be derived in any other way. The common structure of language and the operations of thought are limited to the expression of sequences, whereas functional development grows as a simultaneous process. Word descriptions and functional development are simply not well suited to each other (4).

In forming this chapter, I assumed the reader's familiarity with equation 2.13, the *function of creation*, and with the functional view of the whole process, as described in chapter two (5). With this preliminary background as a given, my formula for the *function of artistic creation* can be introduced without any preparatory discussion. The formula expresses the characteristics of the creative process. After these are described, the formula is used to make further observations about the creative work process, both in art and in other fields. Lastly, we consider the new direction of art, which emerged during the early years of this century, and show how this direction correlates with my formula for the *function of artistic creation*.

1. Art as Creation

Around 1921, Paul Klee wrote that his primary aim as an artist was to come "somewhat closer to the heart of creation than usual" (6). Some years after I read this statement, when my own working method had developed to a state where it was self-sustaining, the relevance of Paul Klee's statement jumped into sharp focus. I recognized it to be a profound observation about the nature of art, a guiding precept that further clarified my own working method.

We can restate this precept more broadly as follows: Original and creative artistic activity is fundamentally a work process that imitates the natural process of creation.

The basic idea of this precept may seem to be simple and familiar — apparently not new — but after years of trying to make its deeper meaning clear to others, I can testify it is rarely understood as an objective statement. Like many of the functional observations in orgonomy, it too requires a drastic change of attitude or a fundamental alteration in one's outlook before it can be understood. Any artist who follows this precept will find it continually modifies his views about the processes of art and directs him toward a new and radical perception of all processes.

In this deeper view, the forming process emerges as having a wider and deeper functional content than the art object or product. As we shall demonstrate, this view is strictly functional, and it coincides with Reich's repeated observations about the priority of the work process in original research. Reich stated that the processes are often more important than the results, and the way of searching supersedes the goals or end products.[2]

2. The Function of Artistic Creation[3]

We begin with the observation that the work process for creating original art imitates the natural process of creation. If this observation is correct, it implies that the operations occurring during the activity of artistic creation are functionally equivalent to the operations occurring in the natural process of creation. Therefore, if we substitute the characteristic functions of the art process in place of the abstract functions in the equation for the *function of creation* (equation 2.13), we should arrive at a functional formulation that specifically describes the process of artistic creation, that is, the *function of artistic creation*:

7.01

[2] The exact quote is reproduced at the head of the preface to this book.

[3] First presented extempore at a seminar on social orgonomy, New York, April 1982.

The **living WORK** function directs the work **CONCEPT** and the work **ACTION**. When the work **CONCEPT** and **ACTION** *fuse* spontaneously, a new creation product emerges. This whole process is recorded in the *structure* of the **WORK OF ART**.

The *work of art*, or art object, represents the new unit or new function, which replaces function **A1** in the general equation for the *function of creation* (equation 2.13). The abstract variations, **Vx** and **Vy**, are replaced by the functional variations of *concept* and *action*. These descriptive terms were chosen because they aptly represent the dissociated functions of intellectual and physical expression, of psyche and soma, of theory and practice, that is, the paired variations that must objectively *fuse* during the creation of an original work of art.

The way we label the paired functions (objectively understood to be two differentiated streams of organismic orgone) is not immediately critical, since it does not affect the essential operation of the *whole* functional process. We would simply replace the chosen terms with better ones, if further studies require we do so.

However, the function that drives the whole process (i.e. the common functioning principle or CFP) must be correctly stated. It has to be capable of generating the paired functions, *concept* and *action*, and of driving the whole process of creating art. We found that the function *living work*, meaning the living effort, correctly represents the function that determines the given variations. This living expression also supports the moving energy that may spontaneously fuse during the process. Though they may be different, the functions *work* and *energy* are simultaneously identical (7). The creation of art is at its root a *work* process. We even refer to the product as a *work* of art. Functionally, we note that *living work* is a secondary manifestation of the life energy, yet it also expresses some organismic orgone functions directly, that is, it expresses some primary manifestations. These observations confirm that the function *living work* is the correct CFP for the process of artistic creation.[4]

3. The Significance of Realms and Domains

Before examining the functional content of our equation, we will briefly review the question of realms and domains of function.

The established way of ranking things in the order of their importance is not adequate for describing the order of functional domains. The functional status of the CFP relative to its functional developments is not a matter of rank or category. The CFP belongs to a *deeper* domain of functioning, so we cannot describe it as having a "higher" rank. Its position on the left of the equation represents a distinction that cannot be understood or analyzed from the viewpoint of its parts or effects, that is, from the domain of the paired functions. The latter belong to a narrower, higher domain. A qualitative and quantitative distinction exists between the two domains that can only be bridged by research, discovery, and proof of a functional connection. The CFP of a known pair of functions cannot be deduced by logic alone — it has to be discovered.

Thus, the horizontal stages in a functional equation separate the functions into functional domains and clarify the order of the domains within a given realm. The CFP on the left, is the broadest, deepest domain, and this is followed by a succession of narrower, higher domains toward the right.

[4] The sequence presented here follows the "historic" order in which the constituent qualities were first ascertained. The order of this kind of research is, inherently, the "reverse" of the direction of development.

4. Functional Observations about Artistic Creation

The equation for the function of artistic creation demonstrates a number of fundamental facts about the creative process:

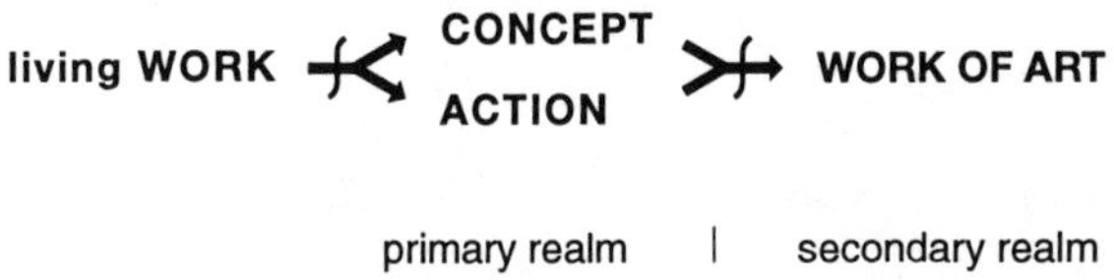

living **WORK** ⤙ **CONCEPT** / **ACTION** ⤚ **WORK OF ART** 7.02

primary realm | secondary realm

- Within this setting, the **WORK OF ART** constitutes the narrowest domain of function;
- The paired functions **CONCEPT** and **ACTION** represent a broader, deeper realm of functioning than the realm of the **WORK OF ART**.

These observations establish, in a new way, the priority of the process over the results or end product. It confirms what I deduced about 30 years ago, and the similar, but prior, conclusions of Klee and Reich, both of whom were experienced in working in unexplored territory.

We must emphasize there is a crucial, qualitative difference between the realm of the *whole* process and the domain of the work of art. This difference tells us we cannot extend our comprehension into the deeper realm while we are completely bound by the narrower realm. *The* work of art *does not determine what we can understand about the deeper realm.* This fact alone, were it understood, could devastate the whole edifice of contemporary art criticism and many of the established practices of art instruction.

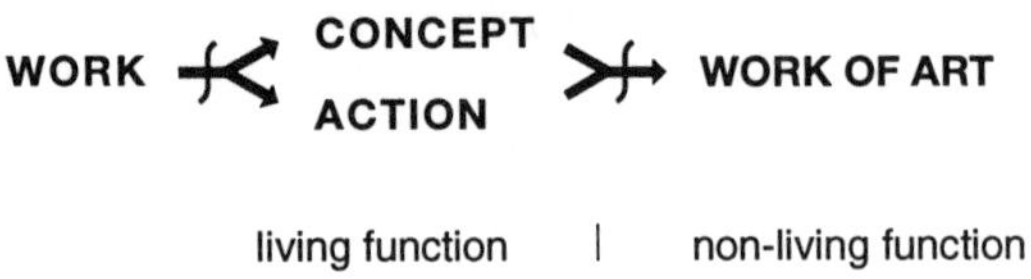

WORK ⤙ **CONCEPT** / **ACTION** ⤚ **WORK OF ART** 7.03

living function | non-living function

- We observe that all the functions to the left of the operation of fusion are living expressions, whereas the new unit, the **WORK OF ART**, is a non-living, frozen structure.

Since art is intrinsically an expression of human life, we would expect to pay more attention to the living function than to the inert product. Besides, the product is never more than a record of what was actually created; no new substance or energy emerges from the product itself. It simply consists of marks or shapes imposed upon existing materials. To recapture the vital expression of a work of art, the viewer must physically reanimate the inert impressions by supplying the movement, just as a tape recording must be moved by a mechanical motor before any sound can be produced.

- We note that the function of artistic creation is also irreversible because the fusion transformation cannot be undone.

Restoring the living functions of *concept* and *action* out of the structuralized substance of the *work of art* does not seem possible.

This correlates with the irreversibility of procreation in life. It would also explain why even the best restorations of artworks turn out to be inadequate and perceptibly anachronistic.[5]

---

[5] Since we repeatedly avoid recognizing that restoration is destructive, many of our treasures are frequently restored. The current careful restoration of Leonardo's *Last Supper*, for example, is bound to alter the quality of what remains, perhaps even exceeding "the ravages of time."

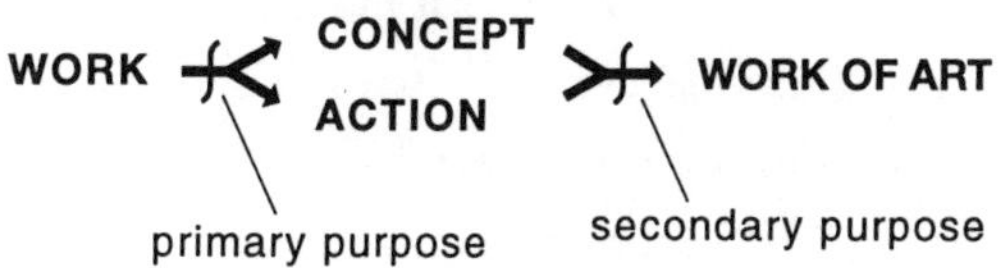

7.04

- We also observe there are two possible locations for the introduction of "purpose" in the total setting. As Reich noted, ". . . function always precedes its purpose and never vice versa," so a primary purpose would occur after the function *living* **WORK** and a secondary purpose after the functions of **CONCEPT** and **ACTION** (8).

It follows that a primary purpose can be used to control the vital functions of *concept* and *action*, whereas a secondary purpose would merely add variety to the end product. For centuries, the practice of art has been directed by specific primary purposes that could only diminish the vitality of the whole process and its product. We are now in a position to see there is only one primary purpose that does not distort the vital functions of artistic creation, namely, the purpose of imitating the natural function of creation.[6]

5. Applying the Function of Artistic Creation

The most difficult problem in applying the function of artistic creation concerns the presence, or relative presence, of natural free-flowing motion as it is expressed through the function of *living work*.

The *function of creation* is determined by natural law (N). In this primary realm, orgone energy moves freely and spontaneously. The imitative CFP, *living work*, should also express the

[6] This correlates with the author's own working precept as an artist.

same freedom and spontaneity as far as possible. But matching this specification is singularly difficult because most of us have blocked the free flow of our organismic energy — the biopathic source of chronic *armor*. The common condition of orgastic impotence, for example, expresses itself through the *living work* function as compulsive work at the one extreme or as an incapacity for work at the other. Both expressions are poor representations of the free, spontaneous movement of function N, though they do reflect its operation in a distorted way.

If the work function is distorted in any way, the whole function of artistic creation will also be distorted, and the distortion will express itself in the *work of art*. Or stated positively, the nearer the imitative process comes to reflecting the spontaneous and free functions of primary orgone, the nearer we can come to the heart of creation.

6. The Impact of Photography on Modern Painting

I like to think that the historic process leading to the radical change in the direction of painting was initiated by the invention of photography (1839).[7] Painters were immediately intrigued and excited by the first photographs. Some of them assimilated photography into their technique of painting. Others saw it as a new branch of painting and became involved with the practice of making photographs, which at the time meant making your own plates, shooting the pictures, processing the plates — everything custom-made.

However, as the technique of photography improved, some painters realized photographs were actually usurping areas of

[7] First demonstrated to the public by Daguerre and Niepce.

the territory that had always belonged to painting. In portraiture, for example, the photograph had a decided advantage over painting — speed of execution, accurate images, and very low cost. Artists were forced to reassess the characteristics of painting and to identify those characteristics that were qualitatively superior to anything photography could accomplish.

The methods of Impressionism and Pointillism developed as an immediate and colorful artistic response to the visual challenge of pre-color photography. Cezanne reacted in a more radical way. He saw the photographic image as a challenge that penetrated to the very roots of artistic practices. Though he was never entirely satisfied with his own achievements, which he freely admitted, his deeper approach initiated a fruitful alternative path, later known as Cubism. A third direction, Expressionism, borrowed techniques from both of the other two approaches to achieve a synthesis, in many respects more abstract than either of its sources. It exalted the individual responses of the artist above everything, except the materials of the art.

7. Kandinski Changes Everything

The development of these directions continued to excite and attract many other artists throughout the first decade of this century — as it still does. We can easily imagine the excited feelings of a serious young artist caught up in the middle of such an artistic revolution. Established ideas about what a painting should look like were being altered almost daily. There were hundreds of new questions to resolve, endless possibilities, and nothing was entirely certain any more.

Wassily Kandinski (1866-1944) began his career during this period. He quickly absorbed its many facets by imitating the works of each direction. He was not satisfied, however, and in

1911 spontaneously produced a group of paintings that established an entirely new foundation for art. His early writings show he was intellectually aware of his achievement, and in his theory of art, he had rapidly consolidated the many observations that emerged from his new perspective.

Kandinski's mature paintings confirm that the artist's *work process* somehow contains the unique principle of art. The product of art is special only because it expressed this *process* and for no other reason. Furthermore, the *process* of art somehow reflects an objective characteristic of Nature, which is more fundamental than any photographic or figurative reference. Kandinski felt art should concern itself exclusively with the "inner essence" of things, the qualities of *energy* and *motion*, and the artist should discard the superficial aim of reproducing the outward appearances of things (9).

8. Quotations from Kandinski

> Externally, each individual graphic or pictorial form is an element. Inwardly, it is not this form itself but, rather, the tension within it, which constitutes the element.
> In fact, no materializing of external forms expresses the content of a work of painting but, rather the forces = tensions within it.
>
> (9:33)

> My task is to point out "organic" relationships between the elements of painting. Even in cases where it is impossible to establish identities, that is, to prove them conclusively. I will indicate their inner relationship by the use of two arrows ⇄. Furthermore, one must not in such cases be deterred by possible mistakes: the truth is not infrequently reached by way of error.
>
> (9:75)

Quite apart from differences in character which are determined by the inner tensions, and quite apart from their processes of creation, the original source of every line remains the same — the force.

(9:82)

On an identity of art and nature/science:

This viewpoint has, until now, become evident only in *abstract* art which has recognized its rights and duties, and which no longer leans upon the external shell of natural phenomena. It should not be replied here that this external shell in "objective" art is put to the service of inner purposes — it remains impossible to incorporate completely the inner of one realm into the outer of another.

(9:104)

And the goal of a theoretic investigation is

1. to find the living,
2. to make its pulsation perceptible, and
3. to determine wherein the living conforms to law.

(9:145)

9. Paul Klee's Expansive Development

Paul Klee was well-prepared for the radical change Kandinski introduced. Starting with the principle of art as a *process*, he began to observe the processes of nature and note the congruences between the two. He even came to think in terms of functions and functioning, always comparing his own working process with natural processes. He developed a functional understanding of energy, motion of expression, expression in motion, and the role of creation. We can demonstrate this with the following quotations from translations of his extensive pedagogic writings (10, 11):

(The artist) . . . does not attach such intense importance to natural form as do so many realist critics, because, for him, these final forms are not the real stuff of the process of natural creation. For he places more value on the powers which do the forming than on the final forms themselves.

(10:45)

The deeper he looks, the more readily he can extend his view from the present to the past, the more deeply he is impressed by the one essential image of creation itself, as Genesis, rather than the image of nature, the finished product.

(10:45)

Then he permits himself the thought that the process of creation can today hardly be complete and he sees the act of world creation stretching from the past to the future. Genesis eternal!

(10:45)

Such mobility of thought on the process of natural creation is good training for creative work.
It has the power to move the artist fundamentally, and since he is himself mobile, he may be relied upon to maintain freedom of development of his own creative methods.

(10:47)

This being so, the artist must be forgiven if he regards the present state of outward appearances in his own particular world as accidentally fixed in time and space. And as altogether inadequate compared with his penetrating vision and intense depth of feeling.

(10:47)

A concept is not thinkable without its opposite. Duality treated as a unity.

(11:15)

The study of creation . . . emphasizes the path to form rather than the form itself.

(10:17)

What we are after is not form but function.

(11:59)

[Today's artist looks] . . . towards the essential, towards the functional as opposed to the impressional.

(11:69)

Movement is the basis of everything. . . . The work of art, too, is first of all genesis; it is never experienced purely as a result.

(11:78)

10. The Impact of this New Direction

Both Kandinski and Klee have extensively influenced the outward appearance of the art of this century. But for the most part, what has been transmitted is very shallow, i.e. functionally narrow.

The outward elements of Paul Klee's images do not, by themselves, lead everyone to an understanding of the fundamental process he developed, namely, the path of *imitating the process of creation*. Techniques can be isolated, learned, and applied without comprehending the whole system of thought and action by which they were created. The influence of Paul Klee has hardly penetrated any deeper than this. It reminds me of how Reich's observations and techniques have been extracted, isolated, and then incorporated into various contemporary therapeutic systems, while the fundamental source or functional core of his discoveries has been fearfully avoided.

The many styles of contemporary art might lead one to conclude that every style has its own validity, but this is not true for our time — a time that follows a simultaneous broadening in our understanding of art and the emergence of a new direction for art. Nor is it true for a time that includes the extensive discoveries of Wilhelm Reich, which, when applied to art, confirm the fundamental observations and radical innovations of Kandinski and Klee.

Summary

According to what I have demonstrated, the primary function of art is the expression of the *function of creation*. This finding consolidates the earlier Kandinski-Klee perception of the primacy of process in art.

With the *function of artistic creation*, we introduce an entirely new base for the practice and study of art. The formula I derived resolves old questions about the source of the artistic impulse, about its depth, and about its intrinsic spontaneity. Given this view, the relevance of the earlier creations of Wassily Kandinski, Paul Klee, and a few others becomes sharply articulated from the confusion expressed in the many variations of Modern Art. And beyond our subject, we can now see the *function of artistic creation* as the common process for all the artifacts and abstractions *created* by individuals in science, as well as in art.

Postscript

The equation for the *function of artistic creation* is the seminal core of the preceding text. Constructing this variation from Reich's original formula was my first practical exercise in applied orgonometry, and the experience rapidly expanded my understanding of the whole technique. I did not choose or determine this pathway to understanding, but was led there by an original, spontaneous insight. Thus, it was the expression of creation — a creation of thought — that propelled me into a deeper study of the least understood formulation in orgonometry, namely, the *function of creation* (equation 2.13).

11. Understanding a Mass-free Orgone Expression

We should not let "a possible confusion of concepts" interfere with understanding this subject, such as the thought: Reich's creation formula describes a natural process, whereas the artistic creation formula describes an artificial, manmade process. Distinctions of this kind are themselves artificial with respect to the *function of creation* (i.e. an error of thought). The *function of creation* is a mass-free orgone expression, entirely self-determined and primary, and therefore not subject to our control. Quite the contrary, it determines us. Therefore, if the *function of artistic creation* is indeed a simple variation of the natural *function of creation*, as we have demonstrated, then its product is likewise determined by mass-free orgone energy. The artist merely participates as the living "vessel" through which creation functions.

Since the *function of artistic creation* takes place within the protoplasm of the living organism (not outside or beyond the organism), we must assume the *objective* product (**A1**) is a new organismic function. I think it is linked to the process of *living memory*, but this is not yet confirmed. Certainly, the *work of art*, an external object, is not the real product of artistic creation but its abstract *representation*. However, I recognize there are special circumstances where the new creation is expressed outside the organism that initiates it, such as in Reich's Experiment XX (which produced flakes of living proto-matter) and in the creation of nuclear reactions (which produces transmutation products) (12).[8]

[8] Review the discussion of nuclear transmutation in the Addendum of chapter three.

8
HEGEL'S DIALECTIC CONCEPT

Sections

1. A Workshop Definition of the Subject to be Investigated
2. The Workshop Definition Compared with a Functional Development
3. Concerning the Paired Relationship
4. Checking for a Possible Structural Contradiction
5. The Development of the Whole Concept
6. Why the Arrangement Appears to be "Reversed"
7. Introducing a Thought Experiment
8. Reconstituting the Whole Equation as a Regular Development
9. Contradictions in the Meaning of Synthesis
10. Assessing the Content of Hegel's Formulation

The most general functioning principle is contained in the smallest, special functioning principle.

WILHELM REICH (1)

I came to review Hegel's dialectic concept out of a curiosity about its role in the early development of Wilhelm Reich's method of thinking.[1] However, the result of the investigation demonstrated that the connection between them is far from direct and more like the evolved difference between two living species. A full presentation of this fascinating discovery will have to wait for another occasion, but here we should note that Reich's mature method of thought is not simply a functional variation of Hegel's dialectic concept.

Aside from the above, this investigation has satisfied my initial curiosity about the connection between Hegel's method and the technique of orgonometry. We learn an important lesson concerning the correct technique for translating a past historic concept into an orgonometric representation. Furthermore, we are led to a worthwhile orgonometric thought experiment that completes our functional grasp of Hegel's method.

The organization of this chapter again reflects the sequence of the actual investigative process. By following this principle, I hope to convey a feeling of the spontaneity essential to the successful use of orgonometry in a new application.

[1] Georg Wilhelm Friedrich Hegel, 1770-1831. German philosopher.

1. A Workshop Definition of the Subject to be Investigated

The basic process of Hegel's dialectic method may be summarized as follows:

Every assertible proposition (*thesis*) can be opposed by an equally assertible yet apparently contradictory proposition (*antithesis*). This mutual contradiction may then be reconciled on a higher level of truth by a third proposition (*synthesis*).[2]

For the purpose of this chapter, I will refer to this description as the "workshop definition."

2. The Workshop Definition Compared with a Functional Development

The dialectic process, as described above, has features reminiscent of the relation of functions in a functional development. Even more, it resembles an orgonometric *thought experiment,* whereby we test the validity of an observed (or conceived) relationship of functions.[3] Three constituents are named, and their relationship is clearly separated into two domains: A pair of variations (*thesis* and *antithesis*), and a single, integrated product (*synthesis*).

Because the *workshop definition* presents its functional constituents in such a recognizable pattern, we immediately anticipate converting the whole statement into an orgonometric development equation. However, we know from past experience with orgonometric analysis that the natural order of domains requires us to examine and understand the paired

---

[2] Adapted, after reading many other descriptions, from a definition of "Hegelian Dialectic" in the *Random House College Dictionary*.

[3] Reich demonstrated this process in a section called "First Rules of Functional Thinking" (3).

relationship first, that is, before we attempt to link the divide that separates the two domains. This requirement should be treated as a simple rule of orgonometric analysis because, in practice, it helps us reach the meaning of the subject more directly and economically.

3. Concerning the Paired Relationship

The association of *thesis* and *antithesis* can be generally represented in the form of an orgonometric equation for simple paired variations (or functions) as follows (2):

THESIS ⨍ **ANTITHESIS** 8.01

The *workshop definition* qualifies the functional relationship of **THESIS** and **ANTITHESIS** as a "mutual contradiction," and their opposite meanings confirm this description. Therefore, the symbol for an operation of *simple* variations, which is used in the general form above (equation 8.01), can be replaced with the specific symbol for the operation of *antagonistic opposite* variations:

THESIS ⟷⨍ **ANTITHESIS** 8.02

This equation states that **THESIS** and **ANTITHESIS** are related as a functional pair and are identical with respect to an unstated CFP. Furthermore, they are functionally homogeneous, and their relationship is antagonistic and opposite.

4. Checking for a Possible Structural Contradiction

Before proceeding, we carefully re-examine the paired equation (8.02) for any contradictory qualities or possible errors of thought. We even consider the etymological relationship of the paired words, temporarily divesting them of their representational meaning. A conflict could well exist between the relationship of the words as units of language (i.e. with respect to their abstract function in the structure of language) and the functional relationship they are meant to represent. Past experience has shown this subtle contradiction can seriously distort an investigation that involves abstract concepts, but it is not yet known if this problem extends to the terms used for concrete functions (4).

In the present case, the word *anti·thesis* is clearly derived from, and related to, the word *thesis*, i.e. they are etymologically (or structurally) homogeneous. We are satisfied both the words and their representational meaning are entirely integrated with the paired expression represented in equation 8.02.

5. The Development of the Whole Concept

Having established the specific functional expression of the paired concepts, *thesis* and *antithesis*, we can now proceed to formulate the whole development, according to the *direction* of the operation, as described in the *workshop definition*:

THESIS
ANTITHESIS ≻⇸ **SYNTHESIS** 8.03

This equation states that the association of the paired concepts, **THESIS** and **ANTITHESIS**, is perceived to be fused in the concept of **SYNTHESIS** (5).

The *workshop definition* clearly says the third proposition, *synthesis*, is derived from the two paired constituents, *thesis* and *antithesis*. The functional meaning of "higher" in "higher level of truth" supports this view and, to a lesser extent, so does the order of the description. Hegel sometimes thought of the *synthesis* as more fundamental. Nevertheless, the above formulation is entirely consistent with the objective way in which *synthesis* was to be derived (i.e. from an understanding of the paired relationship). For these and other reasons that will emerge later, I am confident equation 8.03 correctly represents the dialectic method, as it is described in the *workshop definition*.

6. Why the Arrangement Appears to be "Reversed"

The orgonometric form of the operation in equation 8.03 follows the direction given in the *workshop definition*. Hegel may have intended to describe a process similar to a functional development, but whatever his intent, his method expresses a *fusion* operation. (Compare the orgonometric symbol in equation 8.03 with the symbol for development, which appears below in equation 8.04). According to its derivation, Hegel's concept of the *synthesis* makes it functionally similar to the *new unit* or *product* of the *function of creation*. Thus, his dialectic method actually describes a process of creation in the realm of thought.

Some years ago, when I first converted the dialectical formula into an orgonometric equation (June 1982), I mistakenly constituted it as a development equation. The confusion which followed was most disheartening, since there appeared to be no sharp method for distinguishing the source of the problem from among the many possible sources of error. I had to consider separately and/or simultaneously the investigator (me), the subject being investigated (Hegel's concepts), and the technique of the investigation (orgonometry's capacity to handle concepts). The solution turned out to be very simple: Follow the actual direction of the development of the concepts according to their historic or genetic sequence. Do not force them into a preconceived arrangement.

7. Introducing a Thought Experiment

I am confident equation 8.03 represents the whole thrust of Hegel's dialectic method in a functional setting. However, I do not claim this represents all of the isolated insights Hegel had about his method, some of which were much deeper than the method itself. Thus, I continue to be curious about the consequences of trying to reconstitute it in the form of a regular development equation. Such an experiment might lead to a new insight, expose a hidden error, or simply confirm our present formulation. We need not worry about a possible confusion of concepts, as described previously, because our investigation has proceeded in an orderly manner, according to the spontaneous logic of functions.

8. Reconstituting the Whole Equation as a Regular Development

We can reconstitute the same three functional constituents in the following experimental arrangement:

8.04

This equation attempts to show the concept **SYNTHESIS** as the CFP of the paired variations, **THESIS** and **ANTITHESIS**. However, functional variations can only be determined by an objective function, i.e. a function that operates in the direction ⟶ of development.

The concept *synthesis* is a post-action expression — an expression which functions as a past, structuralized abstraction that looks toward ⟵ the CFP, which is the "reverse" of the direction ⟶ of development. Concepts of this kind contradict the direction of development. As with a previous example (i.e. the concept of the CFP itself), we are required to represent an objective function in the role of a CFP (6). Reverse concepts, such as the common functioning principle, history, evolution, or synthesis, simply cannot determine anything in the direction ⟶ of development.

To rectify equation 8.04, we must find the correct CFP of the whole development. Since the concept "synthesis" does not fit this role, the function of this CFP can be represented with an abstract letter-label, say **A1**. The concept "synthesis" can be included as a dependent qualifier, attached to function **A1**. This amendment sufficiently alters the functional meaning of the whole arrangement to make it valid, i.e. functionally meaningful. The corrected equation would then appear as follows:

A1, a synthesis ⊰ **THESIS** / **ANTITHESIS** 8.05

In this equation, function **A1** develops and determines the two variations, **THESIS** and **ANTITHESIS**. **A1** functions as the CFP of the whole development — or as a Hegelian "synthesis" — insofar as this equation represents Hegel's perception. Thus, whenever Hegel perceived the concept of "synthesis" to be deep and broad, he was actually perceiving the role of the CFP, i.e. function **A1**.

Equation 8.05 also highlights the fact that function **A1** is either unknown or unstated, even though it may be referred to as the "synthesis," or the "CFP." These reference names are structuralized expressions (i.e. "reverse" ⟵ concepts) and neither can be substituted for function **A1**.

This experimental version of the *workshop definition* (equation 8.05) expresses a concrete development of some kind, but until the exact nature of function **A1** (the CFP) is discovered, we cannot comprehend it fully.

9. Contradictions in the Meaning of Synthesis

The two equations 8.03 and 8.05 are technically valid, but what they describe is substantially different. Equation 8.03 describes the origin of the concept *synthesis*, whereas equation 8.05 describes the common origin of the paired concepts, *thesis* and *antithesis*. Though both represent the operation of the dialectic method, only equation 8.03 expresses the whole content of the given *workshop definition*.

As we understand it now, the dialectic method is directed toward a resolution, namely, the *synthesis*, and Hegel chose this term to represent the amalgam of a conclusion. His interest in the origin of the paired concepts was either limited or subordinated to the main thrust of the method. This is evident in two ways: The order of the constituents — *thesis* ⟶ *antithesis* ⟶ *synthesis*; and the limited description of the paired relationship — only an *antagonistic opposite* pair is considered.

The way we resolved equation 8.05 in our thought experiment and the fact that two different but valid orgonometric interpretations can be derived, direct our attention toward a specific, unresolved error of thought.[4] Hegel is unclear about the nature of order, i.e. functional order, and the *workshop definition* conveys this in the phrase "a higher level of truth." This limitation surfaces in equation 8.05, where we discover it

[4] For any *one* specific process, there can be only *one* complete and correct functional formulation.

expressed as an error of thought in which a structuralized concept ⟵ (*synthesis*) is confused with a concrete function ⟶ (**A1**).

Hegel's unresolved logic can be summarized as follows:

- When he conceived *synthesis* to be like the CFP, he was unaware that one cannot arrive at a concept of a deeper function from the knowledge of shallower functions — *the barrier of realms.*
- When he conceived *synthesis* to be a *creation* of thought, he was unaware of the spontaneous and unrelated changes this process introduces — the emergence of a *new functional unit.*

10. Assessing the Content of Hegel's Formulation

In the context of orgonometry, Hegel's dialectic method emerges as *partly right* (7). We have located and defined the principal error of thought — a misconception of natural order — which severely limits the comprehensiveness and the applications of Hegel's method.

On the positive side, we can also see there is a significant content of *right* in Hegel's *partly right* formulation. His procedural idea of *a basic pattern of relationships*, in which two paired constituents are resolved and somehow embraced by a third constituent, continues to be an approximate, general description of the operation of functions. Because it represents function, this *right* content (or pattern) also appears in orgonometry.

9

THE FUNCTION THAT DEFINES THE GOAL

Sections

1. Functional Order Means More than Sequence
2. The Basic Observation and its Source
3. The Example of Orgone Therapy
4. The Example of the Production of Automobiles
5. The Example of the "End" of a Life
6. The Example of a Reverse-directed Thought Sequence
7. The Goal of Original Research
8. Distinct Functional Differences between Synonyms
9. Process and Political Goals

The goal, the use, the purpose are always secondary to what underlies them as their function.

WILHELM REICH (1)

The concept "goal" is more confusing than it is valuable. Though it has a functional meaning, its meaning expresses an abstract operation of thought, which contradicts the direction of functional development. According to the order of functions, the "goal" of a process is not the expression of the process, nor is it functionally related to the completion of the process. We need to understand that the process is the expression of the "goal."

I shall demonstrate that the function referred to as a "goal" does not lie at the "end" of a development, that is, on the right-hand side of a development equation. Those who think of "goals" or "aims" in this way are completely mistaken. A concept with such a meaning does not exist in functions. Life does not develop toward a "goal." For the "goal" (in a meaningful sense) pre-exists the process in every case, and the function it represents drives the whole process that develops. Furthermore, we never reach the "goal" of many processes, because it is present before these processes begin.

1. Functional Order Means More than Sequence

The old ways of thinking follow the monolinear logic of sequence or "historic" succession, while they unknowingly sidestep the bifurcating logic of the hierarchy of functions. As a consequence, *higher* developments are frequently confused with deeper functions and *primitive* qualities are often treated as shallow or limited functions. This is the problem of a thought process detached from reality, and it is repeatedly expressed in the way we apply the concept "goal" (synonymous with "aim" or "end").

We can all slip into this error of thought simply because the everyday concept of a "goal" is genuinely confusing with respect to its correct direction. In some functional processes, the two kinds of order (functional and sequential) are coincident, and the consequence of the whole development is indeed a deeper function.[1] In others, perhaps the majority, the two kinds of order diverge. Generally, sequence can be characterized as *mechanical* and development as *functional*. However, I can now demonstrate the functional meaning of "goal" in an incisive and comprehensive way, according to its location in a development equation, thereby sharpening our understanding of when to apply this concept. The same orgonometric technique also clarifies why we so readily misapply this concept.

2. The Basic Observation and its Source

A simple but startling functional observation initiated the complete unraveling of the subject of this chapter. It was triggered by a *partly right* (i.e. mistaken) statement: "Life's other cosmically rooted functions have goals, and the goal of therapy is the establishment of orgastic potency (2)."[2] This strained statement juxtaposes two oppositely directed processes as though they were simple variations, and it claims each is directed toward a "goal." If we presume that "goal" means the same thing with respect to each of the two pro-

[1] The *life formula* due to Reich describes a process that is both functional and sequential (see chapter thirteen).

[2] *Partly right* is the correct way to view an error (3).

cesses, then the statement confronts us with a sharp, functional contradiction — usually an indication of an error of thought.

After a while, accepting that the quoted statement must be right in some way, I began to consider the meaning of its contradiction. This more contactful approach spontaneously led to the following orgonometric observation: *Goals only appear to exist when we look back* ← *into the past development of a process, that is, when we look toward the CFP of a development.*

3. The Example of Orgone Therapy

However strange this new conclusion may seem, it can easily be confirmed and reinforced with practical examples. Let us begin with the process of orgone therapy, which as we know has a distinct "goal." Since the CFP of all the biopathies determines the process of orgone therapy, the whole process can be illustrated with the equation that describes the basic development of all the biopathies, namely, the development of *armor.*[3]

ORGASTIC IMPOTENCE
ARMOR ⟨
ORGASTIC ANXIETY 9.01

The "goal" of orgone therapy is the elimination of the CFP of all the biopathies, namely, the function **ARMOR**. We can see from this equation that the "goal" actually appears on the left-hand side of the orgonometric representation, a direction (or view) ← that is the opposite of the direction → of development.

[3] This formulation is derived from Reich, and it represents the first stage of the development of *armor* (4).

We should note, when the process of therapy is successful, that the "goal" is achieved at the "end" of the process. This common term of demarcation (i.e. "end") is structurally expressed in the sequence of the therapeutic process, but it does not reflect the functional order of the development of the *armor*, which is the subject of the therapy. As the therapy progresses, the superstructure of the *armor* shrinks as various constituent functions cease to exist. We need to understand that development can never be reversed, but always proceeds in one direction (⟶ toward the right, as we have agreed).

4. The Example of the Production of Automobiles

The function of a factory that manufactures automobiles is determined by the "end" product, the function *automobile*. Once this manmade function is established, it can be described with the following basic development equation:

AUTOMOBILE ⤙ **MOTOR** / **CARRIAGE** 9.02

The formula for the function **AUTOMOBILE** shows that the "goal" of the manufacturing process appears at the left-hand side of the formula. Again, this example confirms that the "goal" is describing the CFP of the whole functional development. It only appears to be at the "end" of the arrangement when we examine the equation from right to left, that is, toward ⟵ the CFP.

The actual process of manufacturing and assembling an automobile is virtually the "reverse" of a complete description of the function of an *automobile*. However, it is the function of the "goal" that determines the *whole* function of a factory, just as it is the function of the "goal" that directs the whole process of orgone therapy.

5. The Example of the "End" of a Life

Before we can be satisfied, we need to consider two more examples of specific processes that seem to develop in a "reverse" order. The first is the biophysical process of dying. The *whole* function of every individual living creature eventually comes to an "end." The end processes continue to express the function of life, even though we can recognize a reflection of it in some physical, non-living processes, such as the decay of stars. In dying, the constituents of the *whole* cease to function in sequence — there is nothing simultaneous about the process. Some organs continue to live and function independently for a short time after death. Individual cells survive for an even longer time. Thus, the disintegration of an organism, both before and after it has died, follows a progression that appears to be "reversing" the arrangement of a development equation. However, this apparent "reversal" is a conceptual illusion.

According to the logic of functions, any alteration in the *whole* function of a constituent must reflect an alteration in the CFP of the *whole* development. The integration of the constituents expresses the integration of the whole function. Thus, a dying creature is no longer the same *whole* function as it was when fully functioning, and again, a dead "creature" is a different *whole* function.

In a passage discussing the same observation about the integrated *whole* function but in another context, Reich wrote: ". . . the functional whole only remains what it is as long as it is governed by the same CFP. A factory may mechanically be the same producing tools or producing shells in wartime; functionally, it has changed when it switched from producing tools to producing shells since the CFP has changed (5)." Our analysis presents the process of dying as a sequence of whole-function changes. Each change may continue its own development, but no actual reversal occurs. Even in death, *living* continues to be the "goal," or CFP.

6. The Example of a Reverse-directed Thought Sequence

The second example of a process that seems to develop in a "reverse" order is represented by one of the two directions of an orgonometric equation. With respect to the agreed direction ⟶ of a functional development, the direction ⟵ toward the CFP is quite detached from reality. However, the coexistence of these two contradictory directions expresses the operation of thought and the way this differs from the development of functions (6). Our facility for abstraction makes us think there is an obvious "end" to a formulation viewed in the "reverse" direction ⟵, even though functional logic informs us that the quality at the "end" is the deeper, broader function, i.e. takes priority in the order of the functions. Clearly, the same quality cannot be both the "end" (or goal) and the "source" (or priority) simultaneously, unless we understand that these contradictory expressions merely represent two opposite interpretations (viewpoints) of one and the same quality.

As an experiment, we can reverse the arrangement of a known development equation to illustrate the logical contradiction of the concept of a "goal" expressed as an "end." The following is a *partly right* reconstruction of the function of *armor* (see equation 9.01 previously):

ORGASTIC IMPOTENCE
ORGASTIC ANXIETY ≻⟶ **ARMOR** 9.03

When we view this arrangement as if it were a form of development, **ARMOR** appears to be the "goal" or "end" of the process. This viewpoint might lead us to either of two non-branching thoughts: "*Armor* is the goal of *orgastic impotence*," or "*Armor* is the goal of *orgastic anxiety*." However, such a conclusion would be functionally meaningless or false, since **ARMOR** is the "goal" of the paired function, that is, of the *whole* development of the complex biopathic structure that modifies our living functions. Even though we have demonstrated that the concept "goal" refers to the CFP, the above arrangement overemphasizes the separate roles of the constituent functions. The reversed direction of this formulation can mislead us into thinking an individual constituent function, on its own, can lead us toward the "goal" — a serious error of thought.

The original arrangement, equation 9.01, correctly represents the role of the CFP. The arrowheads of the symbol clearly show the development proceeds from the CFP, and the function *armor* determines the whole development that follows. We can see and understand, in the order of a development, that only the *whole* of each functional domain can express the *whole* function of the CFP.

7. The Goal of Original Research

Once the CFP of a *whole* development is known, we could describe it as the "goal" of the *whole* development. In the case of natural processes, there is little to be gained by looking at a development in this way (⟵), but in the case of manmade processes, the concept of the "goal" has some usefulness, e.g. the "goal" of orgone therapy and the "goal" of an automobile factory. However, when we are describing or researching a process but do not know the function in the role of the CFP, we cannot be specific about our "goal," or "aim," in a meaningful way. We must simply admit we do not know the "goal." Thus, in original research directed toward the CFP, we can only generalize about reaching a deeper level of understanding.

Nevertheless, research into unknown territories has been permanently changed by Wilhelm Reich's method of functional thinking and his technique of orgonometry. We now have a real framework of primary realm functions and a few simple orgonometric patterns, which illustrate the development of all functions, including those as yet unknown. With this background, we can sharply define the "goal" of research as:

- Directed ⟵ toward an unknown function in the role of the CFP, or
- Directed ⟶ toward the development or discovery of an unknown functional variation of a known CFP.

In the first case, the real "goal" cannot be described, except as an orgonometric abstraction. In the second case, the real "goal" is known, but the "end" product is unknown. (Note how the meanings of the two concepts, "goal" as a CFP and "end" as the termination of a sequence, correctly diverge in this instance.) The contrasting directions indicate that the processes of the two kinds of research are functionally and practically very different, but this interesting subject will be discussed at a later date.

Summary

The concept of a "goal" is an abstraction, derived from the process of functions, and it expresses a direction ⟵ that contradicts the facts of functions. However, within the framework of orgonometry, we have discovered what this abstraction really represents, namely, the CFP of a given development. And like the concept of the CFP, the "goal" is a reverse abstraction that does not function in the role of a CFP unless its specific function is known (or correctly represented).

Since the CFP both determines and limits the functions that develop out of its expression, we come to see our everyday interpretation of the concept "goal" is functionally meaningless, i.e. *false*. According to our finding, the concept "goal" describes a function that is developmentally first, not last, and hierarchically deeper, not higher. We must abandon the erroneous idea of the "goal" as the outcome of a development. The same observations also describe a CFP, and the concept of the "goal" is nothing other than a narrowly understood variant of the orgonomic concept of the *common functioning principle* or CFP.

Addendum

8. Distinct Functional Differences between Synonyms

The subject of this chapter has coincidentally exposed an interesting fact about language, namely, the existence of serious functional differences between synonyms. Synonyms obviously express different nuances of meaning, but can we conclusively decide whether these differences represent separate meanings or simply facets of one and the same meaning? The words "goal," "aim," and "end" are etymologically separate, and this fact reinforces the feeling that they express quite separate meanings — close, perhaps, but separate.

Our study discloses that the three synonyms represent the role of a function by separately describing it as a quality of order, a quality of direction, or as a quality of sequence. "Goal" expresses order; "aim" expresses direction; and "end" expresses sequence. According to their common meaning (a variation of the concept of the CFP), each represents the same function but expressed with a different understanding and, therefore, a different content. "Goal" expresses the functional priority of the CFP correctly. "Aim" describes it as a direction, but we have no way of knowing whether the direction is toward the left ⟵ (correct for the CFP) or toward the right ⟶ (incorrect with respect to the CFP). This is a crucial factor in differentiating what is being represented — a real function or an abstraction that floats above reality. "End" expresses a quality of sequence, which is entirely irrelevant and confusing with respect to the functional meaning represented by "goal." It only seems to apply in the case of "reverse" processes, like orgone therapy, because we confuse functional priority with the order of the technical procedure (a sequence). By adopting the term "end" as a synonym for "goal," our language itself continues to propagate this error of thought.

Out of the above group of synonyms, only the term "goal" narrowly measures up to describing the characteristics of the CFP. The term "aim" requires further qualification to be useful, whereas the term "end" distorts the functional meaning so much, we recommend it be entirely discarded for this use.

9. Process and Political Goals

Social and political planning is frequently directed toward preconceived "goals" or "ends." The well-known subversive saying, "the end justifies the means," is a compelling general example. Having come to understand the functional meaning of the concept "end" (i.e. goal), we can analyze exactly how this message distorts its functional content.

Briefly, we first consider the functional content of the given sentence, which we find is the relationship of the "end" (i.e. the CFP) to the "means" (i.e. the process or variations). Next, we clear away the bias expressed in the verb "justifies" and replace it with a directly operating and functionally familiar verb, "determines." This adjustment produces the following clean functional restatement:

The end determines the means. 9.04

We can see from this restatement that the original sentence contorts the factual relationship by emotively separating the "means" from its CFP with the verb "justifies," thereby further obfuscating the fact that the "means" is the expression of the CFP (represented here by the label "end").

However, our reconstructed sentence 9.04 still contains a structural contradiction. The term "end" is a "reverse" concept (like the concept of the CFP itself), and it does not function correctly in the role of the CFP (or the role of the *subject* of the sentence). To eliminate this fault, we simply reverse the order of the words in the sentence and produce the following correct reconstruction:

The means are determined by the end. 9.05

Or again, translated into a more familiar functional description of the basic relationship:

The means are identical with respect to the end. 9.06

The last two functionally correct restatements of the original make it crystal clear that foul "means" are an expression of a foul "end" or "goal."

We can all be temporarily misled by the logic of the original statement or others like it, if only because it exploits our tendency to view every function or process in isolation. When this tendency is coupled with one or more misconceptions about the way things really work, functionally distorted observations can be mistaken for profound truths.

EXPLORATION

10

HOW TO INTEGRATE AN UNKNOWN FUNCTION

Sections

In research it is of paramount importance to know exactly what you do not know.

WILHELM REICH (1)

Knowing what we do not know begins with what we know. Any *known* function can be experimentally placed within a functional setting or equation.[1] In such a setting, four other constituent functions are seen to be closely related to the *known* function. Even though the four constituent functions may be *unknown*, we can now describe the basic functional relation of any one of these *unknown* constituents with respect to any other, including the *known* function. The symbol for an *unknown* function is introduced. This notation device sharply differentiates what is *known* from what is *unknown*. With it, we can construct an exact picture of what we do not know, and one which provides leading information for the task of exploring the *unknown*.

1. The Symbol for Functions of Unknown Quality

We often need to illustrate the principle of functional development without being specific about the *quality* of every constituent function. In this context, there are times when even the abstraction of a letter-label can be too meaningful, that is, too expressive. For example, function **Y** and function **A** describe a *heterogeneous* difference. This is more than we

[1] This application of orgonometry is an *experiment of thought*.

could know about an *unknown* function. I therefore introduce a pictographic symbol to represent the *unknown* function:

□ **or** □2

The "empty box," or rectangular shape, represents a function that is anticipated but not yet known. When more than one unknown function is required, we shall add a reference number to each symbol as shown above right.

The above symbol should be used sparingly and for no other purpose than to represent an unknown function while it remains essentially unknown, or pending. As a functional label, the open meaning of this symbol makes it technically unique, and therefore we must agree, in this one case, that the objective characteristics of the symbol do not in any way describe the function it represents. For example, function □**2** and function □**6** are not necessarily of the same kind (i.e. homogeneous), though they share the same symbolic representation.

2. Preliminary Considerations

Any orgonometric investigation of the unknown must begin with at least one known or given function, and we will represent this function with the letter-label **K**. Though function **K** is known, the fact that its relation to other functions is unknown implies function **K** is only understood in isolation — a state of understanding that applies to much of the knowledge mankind has accumulated. Once this is observed, it seems obvious every isolated view of a function should be corrected and replaced by an integrated view of the same function. However, this simple conclusion would not take us any further without the necessary tools for correcting the way we think. With orgonometry, we possess both the forms and the technique for integrating a known function within a *whole* functional development, even if all the other functional constituents are unknown.

The term "unknown" describes that which is *not* known. Yet we *do* know something about the unknown, namely, that it too must function, like everything else, according to the *common functioning principle* (CFP) of nature (2). More specifically, we can anticipate any function will be integrated in some way as a constituent of a functional development. Thus, with the aid of the symbol for functions of unknown quality (introduced above), we can proceed to formulate the relation of an isolated quality (e.g. function **K**) within the pattern of a development equation.

Technically, we may proceed in one of two directions, according to the way we perceive the known function **K**. If we think of function **K** as a variation, then we would proceed toward its CFP — the direction ⟵ of the CFP of all nature. Alternatively, if we think of function **K** as a CFP, then we would proceed toward its variations — the direction ⟶ of development. Both procedures are relevant and both can be integrated as one extended formula. Such an arrangement would express the whole development and all of the unknown functions we correctly associate with a single known function.

To clarify what is knowable about the unknown, we will present each of the above procedures and their individual formulations. In this way, we hope to emphasize the different roles of each of the four unknown functions that are so closely associated with any function known in isolation.

3. Working toward the CFP

Let us say, for example, the given function **K** expresses a narrow quality — a detail — such as a symptom. With this understanding, we would justifiably view function **K** as a *variation* of an unknown CFP and expect to formulate it in that context. Or we might decide, for unrelated reasons, to search for the deeper, broader function (i.e. the CFP) that determines function **K**. Both reasons direct us toward the CFP ⟵, and we would express this in the following experimental arrangement:

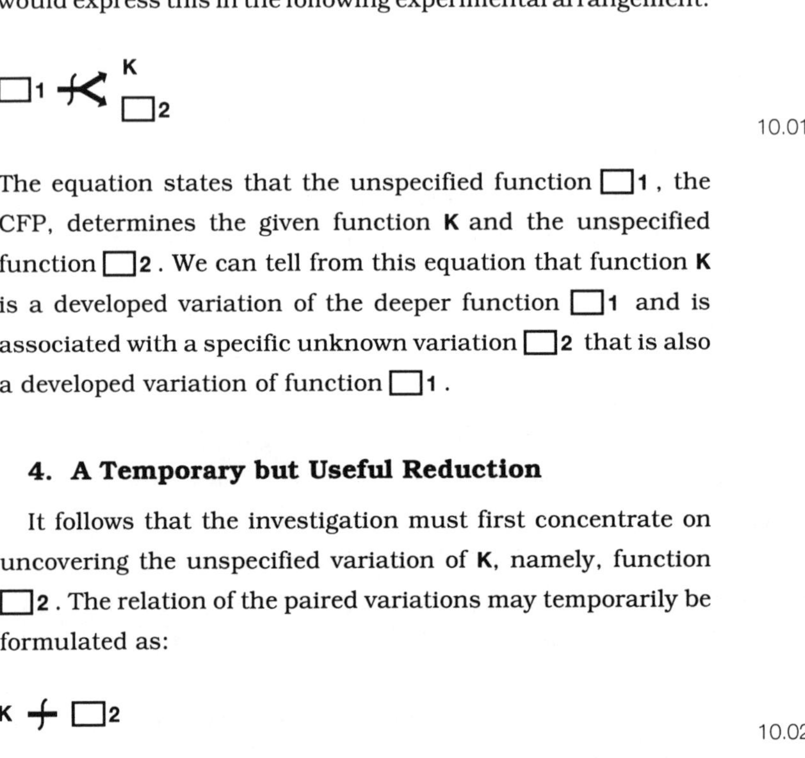

10.01

The equation states that the unspecified function □**1**, the CFP, determines the given function **K** and the unspecified function □**2**. We can tell from this equation that function **K** is a developed variation of the deeper function □**1** and is associated with a specific unknown variation □**2** that is also a developed variation of function □**1**.

4. A Temporary but Useful Reduction

It follows that the investigation must first concentrate on uncovering the unspecified variation of **K**, namely, function □**2**. The relation of the paired variations may temporarily be formulated as:

K ⫟ □**2**

10.02

This equation states that function **K** and the unspecified function □**2** are functionally identical with respect to an unstated CFP (function □**1**, according to equation 10.01).

In this orgonometric form, we are confronted with only one unknown, which would appear to simplify the investigative task. Generally, the search for an unknown function begins as a search for the specific functional variation of a known function. But once we find a possible answer, it can only be confirmed through the discovery of the function that determines both variations and their relation to each other, namely, their CFP.

The present form of equation 10.02 is so generalized we cannot tell whether the two variations are *heterogeneous* or *homogeneous*, nor, if they are homogeneous, how they interact. However, we do know the paired functions share the same domain of function, that is, they belong to the same hierarchy of functioning. And this simple fact can sometimes be very important for the resolution of a functional investigation.

5. Working toward the Variations

If the aim of the investigation is to explore the further development of the given function **K**, or if function **K** appears to express a broad, deep quality, then the first experimental formulation might be arranged in the following way:

$$\mathbf{K} \multimap\!\!< \begin{matrix} \square\,\mathbf{3} \\ \square\,\mathbf{4} \end{matrix} \qquad (10.03)$$

This equation states that the given function **K** determines the unspecified pair of variations, function □**3** and function □**4**. In this setting, function **K** is presented in the role of the CFP, and the unknown paired functions constitute a narrower domain of function.

Since neither variation is specified, we presently have no reason to represent them in the form of paired functions, i.e. as an association of two *unknown* functions.

Original research of the kind represented by equation 10.03 constitutes the bulk of the work accomplished by classical science. We understand, from the development of functions, everything that could develop from a known function operating as a CFP, would be functionally narrower or higher, and therefore, more easily knowable and accessible. However, as the domains become narrower, their overall complexity increases. Classical science has long reached the point where the sheer multiplicity of the separate facts has split its body and thought into a collection of many narrow specializations.

6. An Integrated Representation

A more comprehensive setting for the given function **K** can also be formulated. Equations 10.01 and 10.03 can be simply combined to form the following extended development equation:

$$\square 1 \prec \begin{matrix} \mathbf{K} \prec \begin{matrix} \square 3 \\ \square 4 \end{matrix} \\ \square 2 \end{matrix} \qquad 10.04$$

Here, the given function **K** is seen "embedded" in the *whole* operation of these functions. It is both determined and determining — a variation and a CFP simultaneously. We also see the four unknown functions, □**1**, □**2**, □**3**, and □**4**, which are most closely related to the known function **K**. These constitute the unknowns about which we have some anticipatory functional knowledge.

7. The Unknown within the Whole

The general impression we derive from the above three-stage equation (equation 10.04) emphasizes the integration of the given function **K** with respect to the *whole* development. This forcefully reminds us *no function can be considered in isolation, since no function is in fact isolated.* Similarly, each of the *unknown* functions, e.g. □**2**, is also "embedded" in the same general "matrix" of functional interrelations. By understanding the specifics of this functional "matrix," we can come to understand the roles of functions as yet *unknown*.[2]

Wilhelm Reich began to apply his intuitions and observations of functional order very early in his career. His consistent application of this kind of thinking constitutes the key to the prescience his early work displays. Thanks to Reich and the technique of orgonometry, we too can integrate the *unknown* function within an orgonometric setting and, thereby, exactly anticipate the order of its operations.

Corollary

8. The Known and the Unknown

What we can and cannot know has been the subject of extensive consideration over many centuries. Yet, according to the functional facts demonstrated in this chapter, it has always been possible to know something about everything. This conclusion comes as a surprise, considering the infinity of what we do not know. Nevertheless, were we to recover our original state of *feeling at one with the universe* — in contact with the energy that functions within and without — we would surely find we do know something about everything.

The relation of the *known* and the *unknown* constitutes a functional pair that reminds us of the relation of *right* and *partly right* (3). We can represent this relation as follows:

[2] The terms "embedded" and "matrix" aptly describe characteristics of orgonometric form (*structure*), but are strictly *partly right* with respect to the operation of functions (*development*).

KNOWN ↮ UNKNOWN 10.05

The **KNOWN** and the **UNKNOWN** operate as a pair of antagonistic opposite functions that exclude each other. The two functions are homogeneous, since they are both concepts.

The *known* is defined by the evidence of our senses in combination with the logic of objective functions. Like *right* in the paired relation of *right* and *partly right*, *what we know* and confirm through functional thinking, cannot be whittled away by "pure" logic or arguments based upon such present concepts as the uncertainty principle and/or relativity. Antithetically, nothing is entirely *unknown*, just as nothing is entirely "wrong."

I have demonstrated how knowledge of the process of development, combined with the concept of the common functioning principle, integrates our thinking and makes us recognize that the *unknown* is in reality *partly known*.

11
THE ORDERS OF FUNCTION

Sections

1. The Orders of Function
2. Order and the Generations of the Amoeba
3. Equations with a Mixture of Domains
4. The Development Template: A New Orgonometric Tool
5. A Prepared Template: Custom-made to Explain a Mixed Domain Formula
6. Example of a Mixed Domain Formula
7. The Terminology of Order

It is sufficient in each case where we set up paired functions to be very clear as to whether both functions are on the same or on different functioning levels; the deeper functioning level is always also the wider realm.

WILHELM REICH (1)

The functions of paired variations are narrower than the function of their *common functioning principle* (CFP), and this logic is expressed in natural development as a hierarchy of functions. The logic of this hierarchy is a crucial component of all functional thinking and, therefore, of all functional research. In the orgonometric form of a development equation, this functional hierarchy is visually articulated as a sequence that is read from left to right. Once the technique and the process being represented are understood, we can observe the orders of the functions by simply reading the equation.

The presentation begins with a basic description of the orders of function and their association with the process of functional development. The lack of these orders in our prime example of development — the development of the amoeba — is discussed for the first time. A new orgonometric tool is introduced and then used to demonstrate and explain the meaning and error of *mixed domain* formulations. The chapter concludes with a brief examination of the two technical terms regularly used in describing functional orders.

1. The Orders of Function

Every development equation expresses a hierarchy of constituent functions. This is a practical consequence of making the operational symbol and all the functional constituents follow a given direction of development ⟶ which, by agreement, moves from left to right. The orders that emerge are neither an artifact of the technique nor simply the orders of a sequence but a profound expression of the logic of functions.

The basic form of a development equation expresses these orders as follows:

11.01

| 1 | 2 | domains of function

The numbered scale below the equation marks out the domains, or levels, of function expressed in this development. The scale is merely a convenient reference device, and it has no other significance. Technically, the domains of function are clearly defined by the visual arrangement of the development equation and can be seen by inspection, without the aid of a scale.

Equation 11.01 describes a development in which function **A**, the CFP, develops or determines the paired variations, functions **x** and **y**. Function **A**, the deeper, broader function, represents one domain of functioning (1 on the scale), and functions **x** and **y**, together or individually, represent a narrower, second domain of functioning (2 on the scale). The whole arrangement expresses a functional hierarchy consisting of two domains of function.

This basic hierarchy of function has a profound meaning for correct functional thinking. For example, neither function **x** nor function **y**, nor the combination of functions **x** and **y**, can possibly determine function **A**. Therefore, according to the logic of functions, we cannot even expect to deduce the qualities of function **A** from the limited qualities of functions **x** and/or **y**.

A more extensive development equation shows that the extent of the orders of function merely reflects the extent of the development:

```
                        x
              D ⊀
       B ⊀              y
              E
A ⊀
              F
       C ⊀
              G
```

| 1 | 2 | 3 | 4 | domains of function 11.02

This abstract formula describes a development extending through four domains of functional order. The broadest or deepest is represented by function **A**, the CFP of the whole development. The highest or narrowest domain is represented by functions **x** and **y**, either together or individually. Functions **B** and/or **C** represent the second domain, and functions **D**, **E**, **F**, and **G**, either together, individually, or in any combination represent the third domain of function.

We note, because of the graphic structure of the equation, that the domains appear as vertical groups of related functions. This is entirely consistent with the paired relation of functions, such as **B** ⊹ **C**, **F** ⊹ **G**, or **x** ⊹ **y**, which are vertically adjacent in the arrangement of the equation. The expressive logic of the graphic form of the equation confirms that the operation of any given pair of functions is expressed in only one predetermined domain.

At this point, I would like to introduce briefly an interesting and related technical matter. The orgonometric form for paired variations is generally written horizontally, and this conforms with the notations of algebra and language. However, in a development equation, the paired variations are associated in a vertical lineup, i.e. at right angles to the agreed direction ⟶

of development (as discussed above). This realignment, or transposition, is not simply a matter of notation but also an actual expression of the functional difference between the operations of development and the operations of paired functions.[1] Sometimes, in practice, our understanding of a current experiment or investigation can be improved by simply restating all paired functions in a vertical form — the way they are arranged in a development equation.

2. Order and the Generations of the Amoeba

Functional thinking turns to the living process for its prime examples. This approach was initially intuitive but later was confirmed as the best route to a comprehensive understanding of natural processes. Living functions continue to be used as our prime examples in orgonometry. Hence, when I reintroduced the abstract arrangement of a development equation in "Basic Orgonometry," I related the description to the propagation process of the amoeba (2). The extended version of this development was represented as follows:

A4
A2
A5
A1 → etc.
A6
A3
A7 11.03

| 1 | 2 | 3 | generations

All the functions are letter-labeled **A**, but each letter **A** is individually distinguished by an adjacent small number. This

[1] As far as I know, the operations of paired functions do not express a *development*, though as yet, the possibility of exceptions cannot be ruled out.

kind of *homogeneity* exactly describes the functional relation of one amoeba to another in every generation.

As in previous examples, a scale of functional order is added below the formula, but in this case, it simply demarcates the generations of amoebae. Does this mean that each generation expresses a different order of function? And since the answer is obviously negative, we must ask what makes this prime example of development so different from other expressions of development?

The propagation of life expresses the development of an identical continuity of function — the transmission of a whole living unit of function. This type of development appears to be infinite — it continues to extend without any functional limits. Though the limitations of the CFP limit each constituent function, we cannot say they also limit the continuation of the whole development. For the amoeba, each constituent function is identical to its pair and to its mother or CFP, yet each is unique or different, and some are simultaneous as well. In terms of their function, any amoeba is a simple variation of any other amoeba:

A1 ⨍ A6 11.04

We are led to the functional conclusion that the *whole* development of amoeba (**A1**) is identical to any other individual amoeba (say, **A6**), that is, each constituent amoeba is a simple variation of the CFP **A1**. Thus, the function *amoeba* as a CFP is transmitted *whole* throughout every domain in this living expression of development. Consequently, this basic life process eliminates the usual distinctions of broader and narrower functioning domains. Nevertheless, the development of the amoeba continues to express a sequence of stages, called "generations" — a vestige of the orders of function.

The example of the development of the amoeba reminds us that it is a *function* in the role of a CFP, which determines *development* and which simultaneously determines the orders of other functions. Conversely, we are also reminded that the process of development (a structuralized concept ⟵) does not determine anything.

3. Equations with a Mixture of Domains

Some years ago I noted, whenever I tried to translate certain familiar orgonomic "formulations" into the form of a development equation, technical difficulties appeared that were most confusing. The orgonometric versions seemed to be "closed" and incapable of further development. Yet, whenever the same functional content was reviewed in its original context, each "formulation" returned to being meaningful and functional.

Recently, while working with one such experimental formula, I noted the two given functions of the paired variations were not clearly functions of the same order or domain. Could this be the confusing factor that "closes" the formulation? And continuing with a further question: Could this inconsistency be the common factor in all the other formulations that seemed to be "closed"?

I immediately reviewed a familiar example (due to Reich), which was included in "Basic Orgonometry," namely, *development* paired with *structure* (3, 4). Since *development* is a primary realm expression and *structure* a manifestation of the secondary realm of function, associating the two directly produces a mixture of domains and a *partly right* functional statement.[2] Afterward, I came across more examples of this

[2] There is more than one inconsistency in the original formulation.

kind of mixture, such as the association of *quality* and *quantity*, but reserve discussing these specific mixtures for later.[3] The present aim is to define the mixture of domains and to demonstrate how it can be analyzed and clarified with a newly designed orgonometric aid.

4. The Development Template: A New Orgonometric Tool

I invented a practical tool for demonstrating the functional meaning of a mixture of domains. The basic tool describes the operations and the orders of a functional development without expressing the *quality* of any constituent function. I call this device a *development template* and specify its extent by stating the number of domains. For the present task, we require a *development template* that extends to three domains as follows:

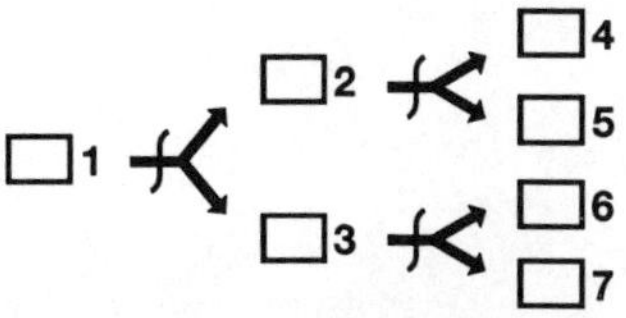

□1 → □2, □3; □2 → □4, □5; □3 → □6, □7 11.05

| 1 | 2 | 3 | domains of function

Since all the constituent functions of a functional development are represented here by the symbol for an *unknown function* □, the whole arrangement consists of nothing more than symbols (5). In this state, therefore, the device is not an equation but rather a pattern or *template* for an orgonometric equation.

[3] Both examples mentioned here will be reviewed in chapter fifteen.

The basic tool or template can now be filled in or adjusted to suit a variety of different tasks. Obviously, the task itself must be generally understood before a specific template can be prepared. For our purpose, the template will be filled in and adjusted to represent the correct relations and functional orders of three constituent functions. The same three functions will later reappear in the *given example* of a mixed domain formula.

5. A Prepared Template: Custom-made to Explain a Mixed Domain Formula

The basic development template above (11.05) can now be reconstructed to illustrate the correct orders and functional relation of three given constituent functions that are letter-labeled **A**, **B**, and **Y**. Labeling the functions transforms what began as a *template* into a development equation as follows:

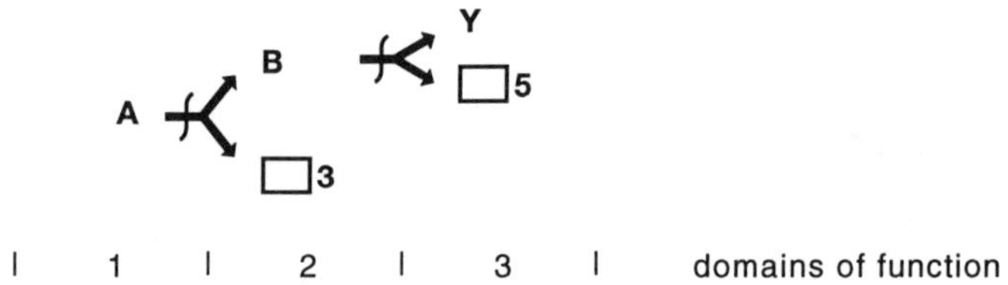

Y
B ☐5
A
☐3

11.06

| 1 | 2 | 3 | domains of function

In this development equation, function **A**, the CFP, determines a paired variation, function **B** and an *unknown function* ☐**3** . In turn, function **B** as a CFP determines the paired variation, function **Y** and another *unknown function* ☐**5** .

Technically, this equation follows the outline of the basic development template shown previously (diagram 11.05). Redundant unknown functions have been trimmed away, thereby altering the immediate form. For easy reference, different letter-labels have been used for each of the known constituent functions, thereby making them *heterogeneous*. However, this example does not imply that mixed domain functions are necessarily *heterogeneous*.

6. Example of a Mixed Domain Formula

Consider the following development equation to be the *given example* of an abstract "mixed domain" formula:

A —< B, Y 11.07

| 1 | 2 | domains of function

First, let us examine this given formula on its own merits. The formula states that function **A**, the CFP, determines a pair of heterogeneous variations, functions **B** and **Y**. It also states that functions **B** and **Y** are identical with respect to function **A**, the CFP. It illustrates a realm of function limited by function **A**, and the development extends from one deeper domain (1 on the scale) into a second higher domain (2 on the scale).

Without further knowledge, we could not possibly know this abstract formulation is inconsistent. However, if a general map were provided of the true development of functions **A**, **B**, and **Y** (i.e. equation 11.06 above), we could check the given formula against the map and discover the error in the given formula. Having prepared such a map, we can now proceed to do this.

Compare the given example, equation 11.07, with the general reference map prepared previously, equation 11.06. Function **A** is indeed the CFP and function **B** is indeed one of the variations that *directly* develops from **A**. However, function **Y** is incorrectly paired with function **B**, since it belongs to a narrower domain of function — a third domain that does not appear in the given example (equation 11.07). Unlike function **B**, the connection between function **A** and function **Y** is *indirect*. According to the reference map, function **Y** is actually a variation that develops from function **B** in its role as a CFP. Therefore, we must conclude that the given example, equation 11.07, represents a mixture of domains, which makes it *partly right* and orgonometrically *false*.

In practice, a basic development equation containing a mixture of domains may still appear to be functionally meaningful with respect to its CFP.[4] Nevertheless, when it is presented as an orgonometric equation, the precision of orgonometry will eventually expose the inconsistency and force us to treat such imprecise equations as *pending* or *partly right*. As yet, there is no special test for this kind of error. The mixtures I have uncovered were found by carefully reviewing the qualities of paired functions, or by trying to extend a formula repeatedly but without any success.

7. The Terminology of Order

Most of the specific terms used for functional descriptions were originally used by Reich in his mature papers on functionalism and orgonometry. Among these are two terms used for discussing the orders of function:

Domain — The term *domain* or *domain of function* is exclusively used to denote any single order of functioning without indicating its rank or relation to any other order. The development of equation 11.02 extends through four orders of function, each of which may be referred to as a *domain* of function. Precisely matched paired variations (by orgonometric standards) operate in a single *domain* of function, as determined by the function of their CFP, which operates in a *directly* related, deeper *domain*.

[4] For example, function **A** continues to be the CFP of constituent functions **B** and **Y** in the given example.

Realm — The term *realm* or *realm of function* also describes an order of function but in a more integrated way. It not only describes a domain of function but also all the domains, or orders of function, which develop from the function that determines the *realm*. The meaning of the term is closely associated with the concept of the CFP, namely, the function that determines the order or rank of any given development or *realm of function*. Referring to equation 11.02 as an example, the *realm of function* **A** describes the *whole* development and/or the deepest order of the whole development. It includes the narrower *realm of function* **B** and **C**, and the narrowest *realm of function* **D**.

After the discovery of orgone energy, Reich began using the term *realm* to differentiate the mass-free functions of orgone (i.e. the primordial or primary functions) from all other functions associated with mass, namely, the physical world according to classical science. He referred to the former as *primary realm* functions and the latter as *secondary realm* functions, and we continue this practice.

Reich's formula for the *function of creation* illustrates some of the important qualitative expressions of *primary realm* function and how they transform into a *secondary realm* function (6):

N ⟶ **Vx** / **Vy** ⟶ **A1** 11.08

primary realm | secondary realm

After a *primary* development, the differentiated orgone energy streams (**Vx** and **Vy**) may spontaneously *fuse* to create a new unit of *secondary realm* function (**A1**).

12

THE PRIMORDIAL UNIVERSE

Before the Beginning of Time

Sections

1. The Generalized Representation
2. Extending the Generalized Representation
3. Introducing a Practical Formula for Primary Development
4. Exploring the Relations of the Constituent Functions
5. Characteristics of the Primordial Realm
6. Explanatory Notes and Further Observations

. . . the human animal has known of the existence of the cosmic original energy ever since he started to write his history.

WILHELM REICH (1)

In "How to Integrate an Unknown Function" (chapter ten), I demonstrated orgonometrically that we can come to *know* the *whole* function without knowing every detail. In support of this conclusion, no other example is more telling than the subject of the "primordial universe." What can we possibly know about primordial function prior to the beginning of creation? Classical science, for example, avoids these subjects deeming them to be both logically and factually unknowable, that is, beyond the scope of direct observation and objective research. However, this chapter will show that we *do* know the *whole* function of the primordial realm, even though we do not know everything about its details.

Our investigation begins with what we know — the primordial function, the process of development, and the development of *mass-free* primary realm functions — and proceeds to define the *primordial realm* with orgonometric formulations. It introduces a new, more specific, orgonometric formula for primary development and concludes with a discussion about a previously unnoticed inconsistency that limits the meaning of the original formula.

The *primordial function* existed before the beginning of what we call the "universe," and it continues to exist. A *primordial development* also occurred before the beginning of creation. Though remote in our time and seemingly "unknowable," the basic functions of this *primordial realm* are essentially the same as those existing and presently described as *mass-free primary functions*. The spontaneous logic of objective functions leads us to this conclusion. By carefully analyzing the *mass-free* expressions of primary function within the context of orgonometric representations of primary development, we can now provide the kind of functional information needed to investigate and describe the characteristics of the "primordial universe."

1. The Generalized Representation

The generalized orgonometric formula for primary development also represents primordial development, that is, the development of primary functions before the beginning of the creation of matter and secondary energy. The basic process is described by the following orgonometric formula due to Reich (2, 3):

$$\mathbf{N} \prec \begin{matrix} \mathbf{Vx} \\ \mathbf{Vy} \end{matrix}$$ 12.01

where **N** represents the primordial function, the CFP of all nature; **V** is the principle of variation; and **x** and **y** are the specific variations.

The equation states that the primordial function **N** develops by dissociating into pairs of differentiated, but nevertheless, related primary functions represented by the abstract generality of the labels **Vx** and **Vy**. Since function **N** determines functions **Vx** and **Vy**, functions **Vx** and **Vy** are identical with respect to their *common functioning principle* (CFP), function **N**.

The differentiated primary functions **Vx** and **Vy** are related as a pair of simple variations:

Vx ⊹ **Vy** 12.02

Because the letter-labels **Vx** and **Vy** are abstract generalizaions, we are forced to generalize about the paired relation of the functions they represent. With these letter-labels, the equation continues to express a generalization. We need to be more specific about the constituent functions and their interaction, before we can be more specific about the meaning of this formula.

However, if we agree that equation 12.01 exclusively represents primary development, then **Vx** and **Vy** specifically represent primary variations. Primary variations are known to be differentiated streams of orgone energy, i.e. simple variations of the primary energy. Accordingly, all primary variations are variations of the same kind of function, which means they are all functionally *homogeneous*. Therefore, all differentiated primary functions are, in fact, related as simple variations.

2. Extending the Generalized Representation

The primary development formula (12.01) describes the basic pattern of the process of development as a *whole*. The functional constituents have been generalized so that one simple formula can represent an extensive process, wherein the primary function (cosmic orgone energy) continually dissociates into pairs of differentiated orgone energy streams. We can illustrate the extension of the primary process with a three-stage development equation as follows:

V3
V1 ⊹<
V4
N ⊹< → **etc.**
V5
V2 ⊹<
V5 12.03

Or, we can illustrate the process with a more extensive operational figure, similar to the one previously presented (4):

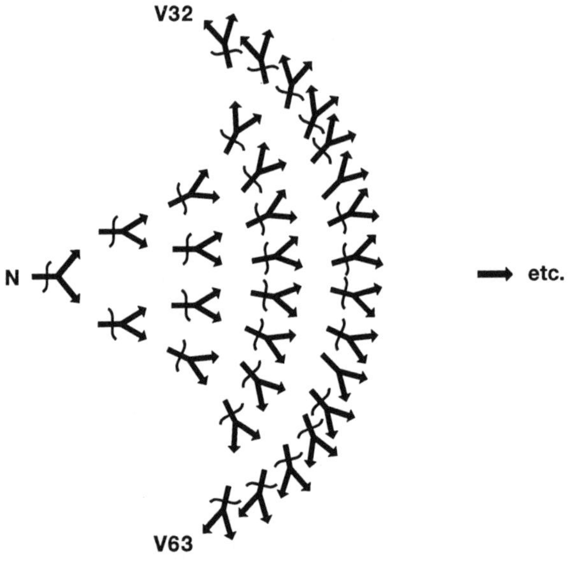

12.04

This depiction of the process of development was originally made to illustrate the infinitely extending operation of the general process. The constituent functions (or their labels) were omitted solely to emphasize the rapidly increasing complexity of the figure as we follow it from left to right.

Neither one of these extended representations conveys the infinite extension of primordial development. Though they adequately describe the pattern of the process, both formulations appear to be describing the "beginning" of development: They start with the primordial function **N** on the left; every constituent function can be easily and directly traced back to the primordial CFP, function **N**; and all constituent reference numbers are small. For some experimental studies, the above representations were found to be too general or too basic, i.e. too close to the "beginning" of everything (function **N**). Other technical alternatives had to be invented to represent *the present, infinite state of primary development.*

3. Introducing a Practical Formula for Primary Development

The above illustration (12.04) is too elaborate to be utilized as a form of equation. A simpler, more practical way to represent *the present, infinite state of primary development* is required, but in a form more specific than the generalization of equation 12.01.

Therefore, I propose using the following orgonometric form to represent an extensive, yet specific, primary development:

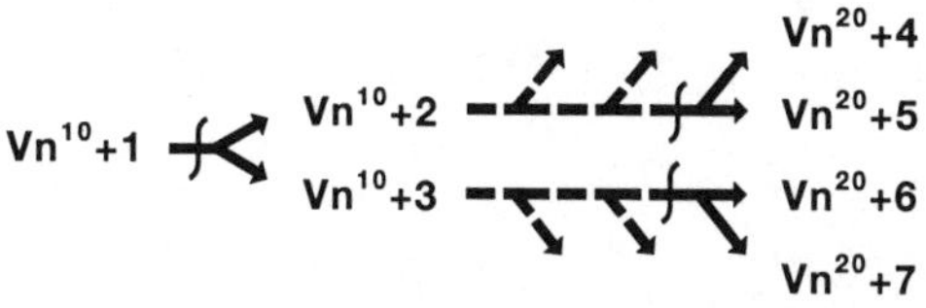

$$V n^{10}+1 \quad V n^{10}+2 \quad V n^{10}+3 \quad V n^{20}+4 \quad V n^{20}+5 \quad V n^{20}+6 \quad V n^{20}+7$$ 12.05

where **V** represents every variation of the primary function; $\mathbf{n^{10}}$ represents an unascertainable but extremely large number; and **+1**, **+2**, etc., distinguishes the individual constituent functions.

This equation shows an extensive process of differentiation or dissociation linking the CFP on the left, $\mathbf{Vn^{10}+1}$, with the remote variations on the right, partially represented by $\mathbf{Vn^{20}+4}$, $\mathbf{Vn^{20}+5}$, $\mathbf{Vn^{20}+6}$, and $\mathbf{Vn^{20}+7}$.

The arrangement is constructed out of orgonometric development symbols linked by broken lines with bifurcations to represent countless stages of development. Each primary variation **(V)** is specified by a reference number stated as a formula (e.g. $\mathbf{n^{10}+2}$). This technical device is used to convey the impression of an extremely large number $(\mathbf{n^{10}})$ and also to provide a practical, individual distinction **(+2)**.

We should note here the CFP of this equation, **$Vn^{10}+1$**, is itself a functional variation whose development is linked, via many stages of primary differentiation (**n^{10}**), to the primordial function (function **N** in equation 12.01). And further, as will be shown, the CFP of the above equation and the primordial function (**N**) are related as simple variations.

4. Exploring the Relations of the Constituent Functions

In the primary realm, each of the constituent functions of a given pair of variations is also a simple variation of every other constituent function. In the above equation (12.05), for example, functions **$Vn^{10}+2$** can be paired with function **$Vn^{20}+5$**, since they are identical with respect to their CFP, function **$Vn^{10}+1$**:

$$Vn^{10}+2 \;\text{ƒ}\; Vn^{20}+5 \qquad 12.06$$

Furthermore, each constituent function in the primary realm is functionally identical to the primordial function itself, that is, the variations express the whole function of their CFP.[1] We could, therefore, illustrate this fact by pairing the CFP of equation 12.05 with either of the constituent functions of equation 12.06 as follows:

$$Vn^{10}+1 \;\text{ƒ}\; Vn^{10}+2 \qquad 12.07a$$

$$Vn^{10}+1 \;\text{ƒ}\; Vn^{20}+5 \qquad 12.07b$$

or by relating all three of these primary functions in two monolinear formulations:

[1] We previously encountered this relation of the CFP to its variations in "Order and the Generations of the Amoeba," which was described as a *transmission of the CFP* (5).

$$\mathbf{Vn^{10}+1 \nrightarrow (Vn^{10}+2 \not\!- Vn^{20}+5)} \tag{12.08a}$$

$$\mathbf{Vn^{10}+1 \not\!- Vn^{10}+2 \not\!- Vn^{20}+5} \tag{12.08b}$$

Equation 12.08a shows that the CFP, function $\mathbf{Vn^{10}+1}$, is qualitatively identical with the combined or individual quality of functions $\mathbf{Vn^{10}+2}$ and $\mathbf{Vn^{20}+5}$. The brackets also isolate the singular function, $\mathbf{Vn^{10}+1}$, thereby formally suggesting this function is technically broader or deeper than the pair of functions contained within the brackets. However, when the equation is viewed by itself (i.e. without any other reference), the functional symbol describing the association of $\mathbf{Vn^{10}+1}$ with the bracketed pair (i.e. a simple paired variation symbol) expresses nothing to support or contradict this suggestion.[2]

Equation 12.08b shows a string of three identical primary variations, which means any one of these can be qualitatively paired with any other. This arrangement emphasizes the qualitative parity or functional identity of all primary variations, including the primordial CFP itself.

In the primordial realm, the *whole* function of the CFP is expressed by every constituent variation in the role of the CFP. Each constituent function determines the same kind of *limit* as the primordial CFP. In practice, the only way we can distinguish between one variation and another is to describe their sphere of operation and/or relative proportions, that is, to articulate them according to the human experience of scale and domain. Thus we perceive some variations to be *cosmic*, others to be *atmospheric*, *organismic*, and *microcellular*. Yet, regardless of their scale, primary variations individually express the *whole* function of the primordial CFP.

[2] In monolinear translations of development, the function in the role of the CFP is often followed by a transformation symbol. When this occurs, the CFP is confirmed by the meaning of this notation.

The above sets of equations, 12.07 and 12.08, demonstrate conclusively that all differentiated primary variations are functionally identical, even though we know primary functions also develop in a sequence of stages. It is quite clear, particularly in the arrangement of equation 12.08b, that the development of primary variations does not express any functional *hierarchy*, which is unlike what was observed about the development of secondary realm functions. Primary functions are entirely *homogeneous*, and therefore, no *transformations* occur in this realm.

5. Characteristics of the Primordial Realm

The creation and development of the technique of orgonometry followed the discovery and early investigations of the properties of orgone energy. This discovery not only confirmed Wilhelm Reich's earlier method of functional thinking but also stimulated the invention of a more precise and comprehensive way to describe the operations of functions, namely, orgonometry. Thus, from its inception, orgonometry incorporated two determining principles confirmed repeatedly by functional logic, observation, and research: First, Nature is fundamentally *unified* by a single primordial function; and second, the primordial function and its primary variations are *everacting*. These two principles are expressed by the orgonometric form of the development of functions, the arrowheads of the development symbol, and the meaning of the direction of development, which is also represented by an arrow. We may note that the idea of an *everpresent* primordial function is conceived out of the abstract fusion of these two principles, i.e. it is a derived, or "secondary," concept.

From the functional information contained in the above formulations, we can extract and list the following basic and general characteristics of the primordial realm:

unified — determined as by a single primordial function **N** (technically inherent);

everacting — continuous action of **N** and all its differentiated constituents (technically inherent);

homogeneous — all variations or constituent functions are of the same kind (indicated by the common letter-label **V**);

equality — all constituent variations are identical with respect to the primordial function **N** (demonstrated by equation 12.07b);

orderly — develops according to a simple functional pattern but without natural hierarchical distinctions (demonstrated by equations 12.01, 12.05, 12.06, and 12.07);

scaling — the pattern of development is identical for all scales, i.e. from the microcellular to the cosmic scale (confirmed by observations of cosmic and microscopic expressions of primary function).

The primordial development is devoid of many familiar secondary realm constraints. These are known from observation, functional logic, and experimentation. Though all of these negative distinctions cannot be extrapolated from the above formulations, a complete list follows:

formless — though primary motion has formal characteristics, it changes continuously and spontaneously;

boundless — the primordial function is expressed as an infinite continuum, and the development of functions (primordial or primary) pervades this continuum;

numberless — the process is infinite, and we cannot even guess the numbers of the distinct functional variations, hence the need for large number references in equation 12.05;

mass-free — mass is an expression of structure (i.e. secondary realm function), which is absent in the primordial realm by definition;

no transformations — *heterogeneous* functions only appear among the products of the *function of creation*;

no hierarchy — primordial variations do not express the functional hierarchy of development (demonstrated by equation 12.08b in conjunction with equation 12.05);

measureless — as we shall prove in chapter fifteen, dimensions do not exist in this realm.

6. Explanatory Notes and Further Observations

In this chapter, the term "primordial" means *before the beginning of creation*. The "primordial function" means the CFP of the *whole* of nature, which is represented as **N** in equations 12.01, 12.03 and 12.04. The "primordial realm" or "primordial universe" refers to mass-free primary functions as they existed before the creation of mass, that is, before the creation of the material universe we inhabit. According to the logic of functions, the primordial realm operated then in much the same way as mass-free primary functions operate now. The presence of secondary structure — matter and secondary energy — does not alter any of the fundamental characteristics of mass-free primary function, and these are the characteristics that express "primordial function." We are sure about this, because we know how to distinguish between *mass-free* expressions and *mass-associated* expressions of orgone.

There are some important functional differences between the *primordial universe* and the *primary realm* as we know it. In this chapter, I excluded any discussion of the *function of creation*, which is a fundamental, mass-free primary process. Though it is a mass-free process, the function of creation generates mass, and this logically excludes it from a description of the primordial realm. Technically, the function of creation "transforms" the "primordial universe" we have described into the existing universe, in which primary realm functions are associated with a secondary realm development.

We do not encounter the primary variations, DOR and Oranur, in the equations that describe the primordial realm.[3] Observations of DOR and Oranur associate these primary variations with secondary chemical reactions and secondary energy radiations. Based upon this evidence, functional logic leads us to conclude that these two variations of primary function (i.e. DOR and Oranur) only develop in the presence of matter and secondary radiations, that is, they never appeared in the development of the primordial realm.

We previously noted that the formula Reich introduced to represent the essence of the process of development, equation 12.01, is technically more generalized than any of his other development equations. By generalizing, he was able to express, with amazing simplicity, the whole process of the development of functions in one formula.

During the composition of this generalized formula, or very soon after writing it experimentally, Reich must have realized it could also represent "primary development" exclusively. As I see it, this practical distinction of meaning was crucial among the steps that led him to formulate the *function of creation*. Here, we can be certain of the sequence of Reich's thoughts, since he technically described the derivation of his groundbreaking formula for the *function of creation*, wherein he uses equation 12.01 exclusively to mean "primary development" instead of "development as a whole" (7, 8).

However, when the generalized equation 12.01 is used to represent primary development, this application highlights an

[3] DOR and Oranur are observed states of atmospheric orgone energy. DOR is *very still* and Oranur is *excited* (6).

otherwise technical inconsistency.[4] By labeling the primordial function with the letter "N" (representing the CFP of all Nature), Reich introduced the orgonometric expression of *heterogeneity* into the general formula for the whole of development. From the evidence presented above, it is clear all the functional constituents of the primary realm are alike, i.e. *homogeneous*. Therefore, to be technically consistent, the constituent functions of an equation for *primary development* should all be labeled with the same letter-label, namely, **V**. This label represents the primary function of all nature as the "primordial variation" (**V1**), a term both meaningful and orgonometrically consistent.

Thus, if we change the label of the CFP, the generalized equation for *primary development* can be correctly reconstituted as follows:

$$\mathbf{V1} \nrightarrow \begin{cases} \mathbf{Vx} \\ \mathbf{Vy} \end{cases} \qquad 12.09$$

where **V1** represents the primordial variation of the whole of nature; **V** is the principle of variation; and **x** and **y** are the specific variations.

The equation states that the *primordial variation*, **V1**, develops the paired variations, **Vx** and **Vy**. Functions **Vx** and **Vy** are, therefore, identical with respect to their CFP, **V1**.

As was shown previously (equation 12.06, 12.07, and 12.08), all primary variations are related as *simple variations*. Thus, the generalizations **Vx** and **Vy**, which represent all differentiated variations of the primary realm are also related as *simple variations* in the following formula:

[4] This generalized formula is significantly inconsistent only when it is exclusively used to describe *primary development*.

$$\mathbf{Vx} \ ⨦ \ \mathbf{Vy} \qquad 12.10$$

The equation states that any primary function, **Vx**, is a simple variation of any other primary function, **Vy**. Logically, the functional content of this formula is the same as the content of our initial "working formula," equation 12.02 previously.

We can also restate the functional content of equation 12.08a and 12.08b in the following generalized form:

$$\mathbf{V1} \ ⨦ \ \mathbf{Vx} \ ⨦ \ \mathbf{Vy} \qquad 12.11$$

The equation shows that all primary variations, **Vx** and **Vy**, are nothing less than *simple variations* of the quality of function **V1**, the singular *primordial variation.*

Having now introduced a new and extended working form for *primary development* (equation 12.05), in which the above technical inconsistency is eliminated, there seems to be no immediate need to correct the historic model (equation 12.01). But the same technical inconsistency reappears in the equation for the *function of creation* (9). In this formula, changing the functional label "N" to "V1" produces a meaningful improvement. This vantage, together with the direction of current orgonometric experiments based on the *function of creation,* may force the issue and make us adopt the label "V1" permanently.

While working on this chapter, I noted spontaneously that the development of primary functions and the mitotic development of living cells are distinctly similar. The individual functional constituents in the development of the amoeba, for example, are all related as simple variations. All amoebae are functionally identical with respect to their CFP, and each is identical to the parent of all amoebae, the CFP of the species.

All amoebae are also *homogeneous*, and their development, like primordial development, does not express its functional order as a hierarchy of functions (10). This kind of functional evidence suggests mitotic development is yet another direct expression of the primary realm as it functions within the living plasma. It certainly supports the orgonomic view that living functions are directly determined by *organismic orgone* (a primary variation) that accumulates and continues to function within the plasma.

13
THE FUNCTION OF THE ORGASM

An Orgonometric Review

Sections

1. The Life Formula
2. The Reference Formula
3. Preliminary Analysis of the Constituent Functions
4. Swelling (Tension) and Relaxation
5. Charge and Discharge
6. Swelling and Charge
7. Discharge and Relaxation
8. Preliminary Analysis of the Deeper Constituent Functions
9. The Operation of Plasmatic Expansion and Convulsion
10. The Function of the Orgasm
11. The Orgonometric Form of the Life Formula
12. The Logic of the Orgastic Cycle and Sequence
13. The Operation of Mechanical Motion and Energy Motion
14. The Reduced Version of the Life Formula
15. The Transformation Symbol and the Operation of a Changeover
16. Terminology

In order to describe a functional realm, it is not sufficient to know the present state of the functions; we must know their functions of transformation if we want to comprehend the whole as a process; otherwise we cannot use the term "functional," and we deviate into mechanistic rigidities.

WILHELM REICH (1)

The profound role of the living *function of the orgasm* and the known facts about its constituent functions were originally discovered by Wilhelm Reich and fully described in a number of major publications (2, 3). My concern has been to restate what we know about this basic living function in the more precise forms of orgonometric equations (4). To do this, I have had to uncover and clarify the exact operations of its constituent functions and find the *common functioning principle* of each pair of constituents. My work on this subject did not proceed quickly or smoothly, but in a series of distinct stages triggered mostly by spontaneous insights. The whole study stretched over many years. Now, having accomplished my objective, I can briefly present the logic of the natural function, its constituent orgonometric equations, and the complete development equation for the *function of the orgasm*.

This review begins with an orgonometric analysis of the functional constituents of Reich's *life formula*. The constituents are restated as orgonometric paired functions and then as development equations. Their common functioning principles are then related as paired functions and again as development equations. One of these development equations is determined by the common functioning principle of the *function of the orgasm*. With this basic function, we can construct a single formula for the whole development of the *function of the orgasm.* We can also reconstruct the *life formula* as a monolinear orgonometric equation, and discover the specific operation that determines the sequence of its functions.

1. The Life Formula

Wilhelm Reich described the sequence of functions that constitute the process of the orgasm, and he represented this sequence in the following way (4):

mechanical TENSION ⟶ bio-energetic CHARGE ⟶ bio-energetic DISCHARGE ⟶ mechanical RELAXATION 13.01a

The constituent functions were derived by direct observation of the qualitative changes expressed in the general process. But the correct order of the constituent functions was not immediately clear. Reich resolved this problem by experimenting with the arrangement of the functions, and by considering the functional logic of each arrangement — a technique he later described as an *experiment of thought* (5). Such thought experiments precede and direct a functional research project and are an essential component of the practical technique of thought known as *orgonomic functionalism* (6).

This early functional formulation proved to be both significant and productive for the development of Reich's work. Not only does it describe the immediate process of the orgasm but it also describes other pulsating life functions, such as peristalsis and cardiac motion. Reich also noted that this kind of pulsating motion, which operates according to the functional sequence above, is exclusively an expression of the living. He concluded the functional sequence expressed in the orgasm was nothing less than the "life formula" itself.

2. The Reference Formula

Though the early version of the *life formula* (13.01a) continues to be a correct statement, Reich later improved the word-label expression of the original formula by replacing the general term "tension" with the more explicit term "swelling" as follows (7):

mechanical SWELLING ⟶ bio-energetic CHARGE ⟶
bio-energetic DISCHARGE ⟶ mechanical RELAXATION 13.01b

This version of the *life formula* will be our starting point and basic reference.[1]

We should note here the above formulas are *not* orgonometric forms. The original was written and published long before the invention of orgonometry. What makes this immediately obvious is the use of standard arrow symbols, whose meaning is less specific than the assigned meanings of orgonometric symbols.

3. Preliminary Analysis of the Constituent Functions

Given the four-function sequence of the *life formula* above, we proceed by considering how its constituent functions may be formulated as paired variations. If this can be done and the related variations are functionally meaningful, they will direct us toward the unstated *common functioning principle* (CFP) that determines the association of each functional pair. This technique will expose a deeper realm of function and, thereby,

---

[1] Formula 13.01b is also known as the *orgasm formula*, but throughout this chapter, I shall refer to it as the *life formula*.

open a path to a more comprehensive understanding of the natural process being investigated. Even if the unstated CFP is an *unknown* quality (a condition often encountered in original research), we can anticipate its existence from the pair of variations it determines.

A preliminary assessment of the four constituent functions of the reference formula (13.01b) suggests, out of the six possible combinations of paired constituents (i.e. according to "pure" logic), there may be as many as four arrangements of paired constituents, which are functionally meaningful (i.e. according to functional logic). Here, we are concerned only with the four meaningful associations. By using the orgonometric symbol for *simple variations*, we can represent these four paired variations in the following temporary way:

mechanical SWELLING ⊹ mechanical RELAXATION 13.02a

bio-energetic CHARGE ⊹ bio-energetic DISCHARGE 13.02b

mechanical SWELLING ⊹ bio-energetic CHARGE 13.02c

bio-energetic DISCHARGE ⊹ mechanical RELAXATION 13.02d

Before each formula is examined on its own, there are a few general observations best made when they are seen together. The paired functions in formulas 13.02a and 13.02b are of the same kind, i.e. *homogeneous*, whereas the paired functions of formulas 13.02c and 13.02d are of different kinds, i.e. *heterogeneous*. We can anticipate, therefore, the functional mix of the latter will be reconciled by a different kind of CFP from that which determines the two *homogeneous* formulas, 13.02a and 13.02b.

SWELLING (or *tension*) and **RELAXATION**, **CHARGE** and **DISCHARGE**, are common expressions of non-living as well as living functions. For example, the functions of **CHARGE** and **DISCHARGE** are constantly utilized in the electrical mechanism of everyday devices; and in nature, the mechanical *tension* (equivalent to **SWELLING**) that builds up along faults in the earth's crust, returns to a temporary state of **RELAXATION** once the tension is released in an earthquake.

4. Swelling (Tension) and Relaxation

In sexual excitation, we see an increase in the turgor of the tissues, especially of the genitals. This is followed by a decrease in the turgor of the tissues, which coincides with the decline of the excitation. The turgor represents a *mechanical tension*, or *swelling*, whereby organs, tissue, and individual cells are filled with fluids ("erection"); the decrease of the turgor corresponds to a *mechanical release* or *relaxation*. The functional relation of this pair of homogeneous variations can be represented as follows:

mechanical SWELLING ⇄ **mechanical RELAXATION** 13.03

Mechanical SWELLING and **mechanical RELAXATION** operate as an alternating pair of functional variations.

The functional alternation of *swelling* and *relaxation* expresses an alternation in the movement of body fluids and a simultaneous alternation in vasomotor activity (8). Thus, the CFP of this alternation is the mechanical *motion* of systemic body fluids, that is, a *bio-mechanical motion*:

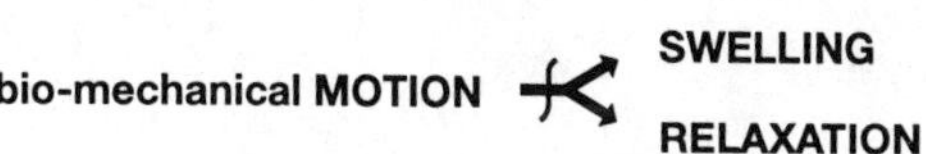

13.04

This development equation shows that a systemic or local **bio-mechanical MOTION** determines the alternation of the paired variations, **SWELLING** and **RELAXATION**. The functions of **SWELLING** and **RELAXATION** are identical with respect to the function of **bio-mechanical MOTION**.

5. Charge and Discharge

The further increase of sexual excitation that follows *swelling* of the tissues is expressed by an increased bio-energetic *charge* at the surface of the organism. After the orgastic acme, the decrease of the excitation is expressed by a decrease or *discharge* of the bio-energy of the organism. The functional relation of this pair of homogeneous variations can be represented as follows:

bio-energetic CHARGE ⇄ **bio-energetic DISCHARGE** 13.05

Bio-energetic CHARGE and **bio-energetic DISCHARGE** operate as an alternating pair of functional variations.

The functional alternation of *charge* and *discharge* expresses the alternating movement of energy within the whole organism. During the phase of excitation, energy moves to the periphery of the organism where the *charge* can be demonstrated and measured (9, 10). After the *discharge*, the diminution of the peripheral *charge* can, likewise, be demonstrated. The movement of *bio-energy* governs the functions of *charge* and *discharge* as represented in the following development equation:

13.06

This equation shows that a systemic or localized **bio-energy MOTION** determines the alternation of the paired variations, **CHARGE** and **DISCHARGE**. The functions of **CHARGE** and **DISCHARGE** are identical with respect to the function of **bio-energy MOTION**.

6. Swelling and Charge

The mechanical *swelling* or turgor of the tissues, especially of the genitals, is followed by a rise in excitation and an increase in the bio-energetic *charge*. The *charge* is at the periphery of the organism, but it does not arise at the periphery without the prior *swelling* of the peripheral organ.

Since *swelling* and *charge* are heterogeneous functions, they are directly related by the operation of a transformation, whereby the one function is changed into the other function as follows:

SWELLING $\nrightarrow$ **CHARGE** 13.07

The peripheral **SWELLING** transforms into a bio-energetic **CHARGE** at the periphery of the organism.

According to the logic of functions, the mechanical *swelling* (or tension) must precede the bio-energetic *charge* because the converse is functionally meaningless. Reich's observations and clinical experience of pathological functioning confirmed this order (11). Thus, the above transformation only occurs in one direction — from *swelling* toward *charge*.

The *charge* at the periphery originates from an energetic excitation in the center of the organism, which then moves from the center to the periphery — an *expansive* orgonotic streaming. These facts are supported by microscopic observations of the process of peripheral extension in the amoeba. Prior to forming a pseudopod at its periphery, bions luminate in the center of the amoeba and a streaming of its plasma sets in from *center to periphery* (12). These plasmatic currents are common to all organisms.

The mechanical *swelling* and the bio-energetic *charge* at the periphery of the organism express the deeper function of the *expansive* orgonotic streaming. This development can be formulated as follows:

SWELLING
EXPANSION ⪪
CHARGE 13.08

The equation shows that the paired variations, mechanical **SWELLING** and bio-energetic **CHARGE** are identical with respect to the function of **EXPANSION**. The **EXPANSION** determines both the **SWELLING** and the **CHARGE** at the periphery of the organism.

The orgonomic term "expansion" represents the spontaneous outward movement of *orgonotic streaming* in a living organism, that is, outward from the functioning center of the organism to its periphery. In classical physics and engineering, the term "expansion" generally represents the response of a *structure* to changes in temperature and/or pressure. With the above equation, we see that this change in *structure* is conceptually related to the energy (secondary) expressed in the *motion* of atomic particles, atoms, and molecules, or on a cosmic scale, in the *motion* of cosmic bodies. (Although these two meanings for the term "expansion" seem to be poles apart, there is a chance, on a deeper level of function, that the functions they represent may well be profoundly similar — as their common name implies.)

7. Discharge and Relaxation

The orgasm is followed by a rapid bio-energetic *discharge* experienced as a steep "drop" in the excitation of the organism. As the discharge subsides, a general mechanical *relaxation* spreads through the whole organism. The feeling of gratification coincides with the mechanical *relaxation*.

Since *discharge* and *relaxation* are heterogeneous functions, they are directly related by the operation of a transformation, whereby the one function is changed into the other function as follows:

DISCHARGE ⇸ RELAXATION 13.09

The bio-energetic **DISCHARGE** transforms into a mechanical **RELAXATION** of the whole organism.

According to the logic of functions, bio-energetic *discharge* must precede the mechanical *relaxation*. The reverse of this order, a "pure" abstraction, is functionally meaningless. This logical conclusion is confirmed by experience and by observations of the process. Thus, the above transformation only occurs in one direction — from *discharge* toward *relaxation*.

Following the phase of involuntary muscle contractions and an automatic increase in excitation, repeated *involuntary body convulsions* appear in both organisms (i.e. male and female functioning as a single, combined orgonotic system). These convulsions coincide with a further and sudden steep increase of *charge* that rises to the acme. The involuntary convulsions continue through the initial *discharge* of the orgasm and the subsequent steep energetic *discharge* discussed above. The function of these rhythmic convulsions of the total organism dissipates the excitation and the *charge*, thereby producing a general *relaxation* of the organism. This whole functional development can be formulated as follows:

CONVULSION ⤙ **DISCHARGE** / **RELAXATION** 13.10

The function of the involuntary orgastic **CONVULSION** determines the paired functions of bio-energetic **DISCHARGE** and mechanical **RELAXATION**. **DISCHARGE** and **RELAXATION** are identical with respect to the deeper living function of the rhythmic, involuntary **CONVULSION**.

8. Preliminary Analysis of the Deeper Constituent Functions

The functions that determine the four development equations previously discussed (i.e. the CFPs of equations 13.04, 13.06, 13.08, and 13.10), can also be related as paired variations.[2] A preliminary assessment of these four functions suggests there may be two meaningful arrangements (i.e. as paired variation equations), which need to be examined. By using the orgonometric symbol for *simple variations*, we can represent the two paired variations of this deeper domain in the following temporary way:

bio-mechanical MOTION ⊹ **bio-energy MOTION** 13.11a

EXPANSION ⊹ **CONVULSION** 13.11b

According to the logic of functions, there can be only one correct arrangement of paired functions that will represent the function of the orgasm in this deeper domain. Therefore, if the constituent functions are correctly stated, one of the above equations must be redundant. First, we need to consider whether the individual constituents specifically represent liv-

[2] This orgonometric technique will simply confirm the correctness of the functions as stated, but only if they are indeed correct.

ing functions, and second, whether either of the given pairs of function could develop the sequence of the *life formula.*

The functions of equation 13.11a represent secondary variations of *motion* common to both living and non-living processes, namely, *mechanical motion* and *energy motion.* The qualifying label "bio" does not alter this fact. It is also clear (from a preliminary assessment) that the further development of these functions will not generate the specific constituent sequence of the *life formula* (13.01b). For our purpose, there is no need to continue with, or complete, this formula, though we will do so later to demonstrate it is redundant (see Addendum).

The paired functions of equation 13.11b, *expansion* and *convulsion,* appear to be exactly right. Not only are they specific expressions of the living but also specific to the process of the orgasm. The further development of this functional pair generates the correct associations of constituent functions expressed in the *life formula* (13.01b). We therefore continue our effort to resolve and understand the functional realm of this paired relationship.

9. The Operation of Plasmatic Expansion and Convulsion

The operation of orgastic expansion and convulsion is periodic but irregular. The *convulsion* follows the *expansion* and the *expansion* follows the *convulsion.* Although these two functions represent opposite qualities of living plasmatic motion, they are essentially the same kind of function, that is, *homogeneous.*[3] Therefore, according to the logic of functions,

[3] The word-labels *expansion* and *convulsion* (or *contraction*) express a *heterogeneous* word origin, which unfortunately contradicts the *homogeneous* quality of the functions they represent (14).

they relate to each other as *alternating opposites*, and this can be represented as follows (13):

EXPANSION $\rightleftarrows$ **CONVULSION** 13.12

The equation presents **EXPANSION** and **CONVULSION** as a pair of *alternating opposite* functions. The general plasmatic **EXPANSION** of the whole organism alternates with a rhythmic, repeating plasmatic **CONVULSION** of the whole organism. This equation describes the same process represented by the *life formula* (equation 13.01b) but in a deeper domain of function. Technically, it too can be called an "orgasm formula."

The alternation of *expansion* and *convulsion* represents the outwardly directed or inwardly directed *orgonotic streamings* of a whole organism. The function that determines these integrated *streamings* must logically be more primary — an orgone-*physical* function. The accumulating property of orgone (i.e. the reverse of entropy) is responsible for the accumulated energy and vitality that distinguishes living matter from the non-living. This process of accumulation also generates the cyclic drive of the function of the orgasm. Therefore, the function of the orgasm is a direct expression of organismic orgone. In species that have developed two sexes, the function of the orgasm involves superimposition and fusion, wherein the energy fields of two individuals of opposite sex combine and function as a single field. In sexually undifferentiated organisms, the function of the orgasm involves the single energy field of the individual.

It is clear that the functions of *expansion* and *convulsion* are determined by *movements* of organismic orgone. However, the description "orgone movement" is too general to represent the orgasm function definitively. It is also clear that the complete function of the orgasm is a variation of *living pulsation* — a variation that specifically moves the organism *as a whole.* Therefore, I propose the functional label *whole organismic orgone pulsation* to represent the specific CFP of the *function of the orgasm:*

whole organismic ORGONE PULSATION → **EXPANSION** / **CONVULSION** 13.13

A **whole organismic ORGONE PULSATION** determines the plasmatic **EXPANSION** and **CONVULSION** of the function of the orgasm. The functions of **EXPANSION** and **CONVULSION** are identical with respect to the deeper function of **whole organismic ORGONE PULSATION.**

10. The Function of the Orgasm

A comprehensive orgonometric equation for the *function of the orgasm* can now be constructed by integrating equations 13.13, 13.10, and 13.08 in one arrangement as follows:

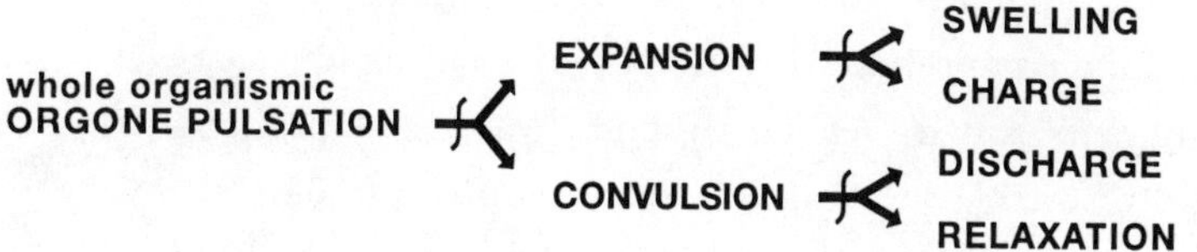

13.14

The equation describes *the function of the orgasm* as a development that extends through three domains of function. The CFP on the left, **whole organismic ORGONE PULSATION**, determines the entire development of the orgasm process. In the second domain, the primary **PULSATION** generates the alternating plasmatic movements of **EXPANSION** and **CONVULSION**. In their turn, the paired functions **EXPANSION** and **CONVULSION** govern the development of the four constituents of the third domain: **SWELLING**, **CHARGE**, **DISCHARGE**, and **RELAXATION**. Thus, the functional constituents of Reich's last restatement of the "orgasm formula" appear on the right of the equation, but here they are arranged in a vertical column that must be read from top to bottom.

The equation for the function of the orgasm sharply distinguishes the three domains of function that express the whole process. The first domain represents the primary function as an orgone-*physical* quality. The second domain represents the direct transformation of this primary operation into integrated movements of the living plasma, i.e. the "living" quality of living matter. The third domain represents the transformation of the general plasmatic movements into secondary living functions, broadly distinguished as "bio-mechanical" and "bio-energetic."

11. The Orgonometric Form of the Life Formula

The form of an orgonometric development equation does not show, as a rule, the detailed interactions of related constituents in any one domain of function. Yet these functional details are crucial for a comprehensive representation of any development of functions. For this reason, the paired variation equations 13.03, 13.05, 13.07, 13.09, and 13.11 are required to complete the orgonometric representation of the whole *function of the orgasm.*

We can condense the functional information given in the first four equations listed above by integrating them in one monolinear orgonometric arrangement as follows (15):

(SWELLING ⇸ CHARGE) ⇹ (DISCHARGE ⇸ RELAXATION) 13.15

Reading from left to right, the equation shows the four constituent functions, **SWELLING**, **CHARGE**, **DISCHARGE**, and **RELAXATION**, in the same sequence as they appear in the *life formula* (13.01b). However, the specific orgonometric symbols and the parentheses add a further depth of functional detail absent in the original formulation.[4]

The parentheses divide the process into two distinct sections linked by the symbol for an alternating operation — the paired variations on the left alternate with the paired variations on the right. The combination of this alternating operation and the consistent one-way (i.e. directional) operation of the two functional transformations within the parentheses, generates the sequence and the cyclic expression of the four-beat process. The above equation (13.15) shows the sequence as a progression that reads from left to right. Thereafter, the process returns to begin again on the left, as indicated by the return arrow ↚, or transformation symbol, which is incorporated in the orgonometric symbol for the operation of *alternating opposite* functions.

Equation 13.15 can be further reduced to show how its constituent functions operate within the limit of one domain of function. In this form, the equation looks like Reich's original *life formula*. However, it is omitted here because it excludes all the deeper functional information present in equation 13.15 (see Addendum).

12. The Logic of the Orgastic Cycle and Sequence

The origin of the *cycle* and the source of the *sequence* of the *life formula* (equation 13.15) can now be more graphically illustrated with the following well-integrated, orgonometric construction:

---

[4] Reich's original version of the *life formula* is not an orgonometric equation. It was published more than a decade before the invention of orgonometry.

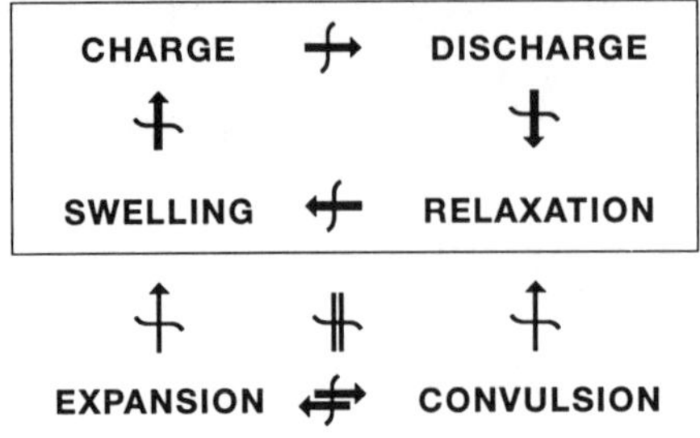

13.16

This orgonometric diagram shows the four-function equation of the *life formula* rearranged in a rectangular "ring" within a rectangular box. We see this "ring" of functions continues to express the correct functional sequence and meaning of the *life formula*. The sequence is defined by the direction of the arrows, i.e. clockwise.

Below the "ring" and the box, the diagram also shows the deeper, alternating operation of the functions **EXPANSION** and **CONVULSION**. **EXPANSION** is vertically aligned with **SWELLING** and **CHARGE**, and **CONVULSION** is vertically aligned with **DISCHARGE** and **RELAXATION**. This vertical alignment expresses the development determined by the functions of **EXPANSION** and **CONVULSION**.

Within the box, the two horizontal arrows (transformation symbols) that link the constituent functions from left to right (above), and from right to left (below), are functionally equivalent to the two directions of the orgonometric symbol for alternating functions.[5] This alternation is determined by the paired operation of **EXPANSION** and **CONVULSION**, the deeper level expression of the *life formula*.

To show the extraordinary "elasticity" of orgonometric notation, I have added the operational symbols that link the deeper

---

[5] In this case, each arrow links a *homogeneous* pair of functions. Therefore, these two arrows represent a simple *changeover* operation rather than a *transformation* operation — the usual meaning of this orgonometric symbol (see Addendum).

formula below with the variations of the *life formula* above. These symbols are directed vertically.[6] Consequently, three more formulations emerge with this arrangement:

- A monolinear equation for the development of **EXPANSION** (vertical left);
- A monolinear equation for the development of **CONVULSION** (vertical right);
- A monolinear equation that equates the orgonometric symbol for alternating opposites with two seemingly unrelated transformation symbols of opposite direction (vertical center).[7]

These three equations are functionally meaningful, and together, they confirm the technical validity of the above diagram.

We learn from the above diagram that the *alternating* operation of the deeper functions, *expansion* and *convulsion*, determines the *cyclic operation* of the four-beat *life formula*. The sequence or *direction of the transformations* of the four constituents is determined by the combined operation of two levels of function: The *alternating* operation on the deeper level, and the operation of one-way *transformations* on the higher level.[8] Thus, in accordance with the orders of function, the expression of the sequence of the cycle is more complex than the expression of the source of the cycle. These conclusions lead logically to a further thought: All cyclic processes, such as the heart beat (living function) or the orbit of a planet (non-living function), may also be governed by a similar, underlying operation of *alternating opposite* functions. Though

[6] For visual clarity, these symbols are thinner.

[7] The constituents of this equation are all *operations* and, as is shown, can be entirely represented with the orgonometric symbols for such operations.

[8] Pathology can reduce or block the constituents of the process but not the direction of the cycle (e.g. priapism, fibrillation, premature ejaculation, intestinal spasm).

not yet confirmed orgonometrically, this thought supports Reich's prior observation of the same deep connection between these seemingly diverse functions.

Summary

By applying the technique of orgonometry to the functional facts uncovered by Reich, we succeeded in restating the *function of the orgasm* in a qualitatively exact, orgonometric development equation. To reach this end, we first had to resolve and formulate the associations and development of all the functional constituents represented in the orgasm process. Once this was accomplished, we could return to restate Reich's *life formula* in the form of an orgonometric equation. For me, the final key to the integration of the whole process was a spontaneous and seemingly simultaneous realization:

- The natural process represented in the *life formula* is actually a *cyclic* process — a quality not expressed by Reich's original formula;
- The four-function *sequence* is the narrower (or higher) level development of a more fundamental *alternating operation* expressed by a broader (or deeper) level of function.

These new observations resolved most of the technical problems I had encountered in my early attempts at formulating some of the above constituent functions. Moreover, the new realizations seemed to lead me directly toward the correct orgonometric representation of the *life formula*. Indeed, realizing how a functional alternation ("two-beat") can develop into a cyclic "four-beat" sequence may be an essential step toward a correct understanding of the functional transformations of the *life formula*.

Addendum

13. The Operation of Mechanical Motion and Energy Motion

As we stated previously (section 8), the paired variations, *bio-mechanical motion* and *bio-energetic motion* (equation 13.11a), are not as specific to living function as their functional labels imply. To emphasize this non-specificity, I will refer to them now as *mechanical motion* and *energy motion*. With this change, their functional development can be represented more generally as follows:

primary MOTION ⤙ **mechanical MOTION** / **energy MOTION** 13.17

The equation shows that the variations **mechanical MOTION** and **energy MOTION** are identical with respect to **primary MOTION**, i.e. *orgone motion*. This formulation holds true for the development of both living, as well as non-living, *secondary* variations of motion.

The further development of the above equation integrates equations 13.04 and 13.06 previously as follows:

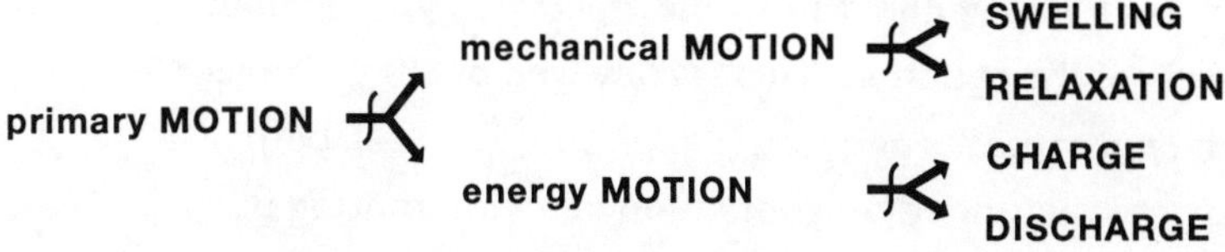

13.18

This development equation shows that the primary function, **primary MOTION**, governs the two secondary expressions of motion, **mechanical MOTION** and **energy MOTION**. These two functions, in turn, determine the paired variations **SWELLING** and **RELAXATION**, and the pair **CHARGE** and **DISCHARGE**.

The narrowest domain of function (vertical right) does not present its constituent functions in the sequence of the *life formula*. Even if we try to rearrange the vertical order of the paired functions (there are only four possible permutations), we cannot achieve the required sequence of constituents in the context of this development formula.

The vertical sequence of the four constituents, which appears on the right of this formula, is merely a formal relationship without any functional significance. In this domain of function, there is no direct operational connection between the paired variations *swelling* and *relaxation* and the pair *charge* and *discharge*. We are forced to conclude that this functional development (equations 13.17 and 13.18) does not generate the *life formula*, nor does it express anything specific to the life process. Furthermore, these paired operations serve to emphasize the correctness and specificity of equation 13.14.

14. The Reduced Version of the Life Formula

Technically, equation 13.15 can be more narrowly reconstructed to show how its constituent functions operate within the limit of a single domain of function. To do this, we must remove all notational references to the operation of deeper functions that determine the cyclic operation of the four constituents of the equation. This is conveyed by the arrangement of the *parentheses* and by the *alternating symbol*, both of which express the underlying operation of an alternating pair of functions. In the domain of the four constituents, the deeper alternating operation is expressed as two changeover operations, represented here by two transformation symbols in the following limited arrangement:

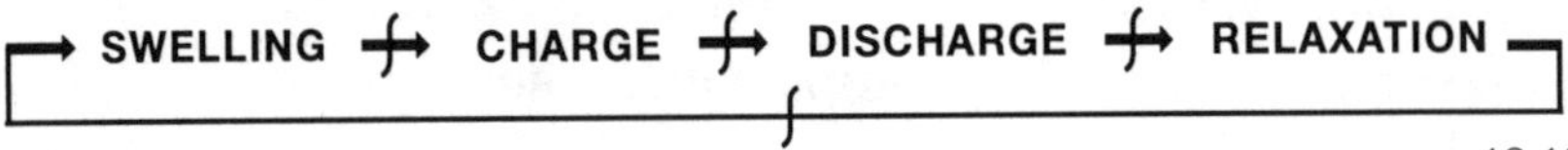

→ SWELLING ⨍→ CHARGE ⨍→ DISCHARGE ⨍→ RELAXATION →
(with a return line from RELAXATION back to SWELLING, marked ⨍)

13.19

This cyclic equation is the orgonometric equivalent of Reich's original *life formula* (formula 13.01b). The one difference we see is that the function of **SWELLING** (or *tension*) follows the function of **RELAXATION** — a *changeover* operation that does not appear in the original formula.

15. The Transformation Symbol and the Operation of a Changeover

The orgonometric symbol for the operation of a transformation has a specific meaning, but it also has a more general meaning (16). Within this general meaning, we now include the operation of a one-way *changeover* that occurs between homogeneous functions of an *alternating opposite* pair of variations (see diagram 13.16). Presently, we can distinguish this other meaning only within the context of an orgonometric equation. Perhaps further work will lead us to create a special symbol for this one-way *changeover* operation.

16. Terminology

Throughout the preceding text, I took great care to use the terms *alternation*, *cycle*, and *sequence*, according to a precise functional view of their meaning. Although my usage conforms with dictionary definitions, two of these terms, *alternation* and *cycle*, can be more precisely defined by orgonometric equations than by word definitions, but this technique will be discussed at a later date.

Sequence — A succession without expressing anything about the quality or hierarchy of any constituent in the succession. With respect to function, this term has no special significance.

Alternation — One of the four basic operations of related pairs of homogeneous functions (17). This operation is generally perceived to be rhythmic, but it can also appear to be aperiodic. It is the deepest expression of all *cyclic* operations.

Cycle — A repetitive operation that develops from an alternating pair of functions. It is expressed as a sequence of functions in an *alternating succession of heterogeneous and homogeneous associations*. The operation of the cycle is uniformly directed by one-way transformations that link each constituent function to the following function. Accordingly, at least two related pairs of heterogeneous functions are required to constitute a *cycle*.

Reich repeatedly emphasized the importance of the *sequence* of the given constituents of the *life formula*. Although I was convinced by the evidence he presented, I also knew that the concept "sequence" has very little functional significance. I was left, therefore, with an unresolved question: What makes the "sequence" so important in this one case? Looking back now, I can see how this "loose end" was a constant stimulus throughout the investigation that concluded with this chapter.

14

THE SOURCE OF TIME AND LENGTH

Sections

1. Summary of Previous Observations and Conclusions
2. First Formulation of the Development of Time and Length
3. Dimensions Also Express an Order of Function
4. Are Time and Length Primary Expressions?
5. The Error of the "Fourth Dimension"
6. Past, Present, and Future
7. Observations that Contradict the "Arrow of Time"
8. Time is Not a Motion
9. Thoughts about Errors Led to the Answer
10. A More Accurate Development Formula
11. The Source of Time and Length — the Whole Formula
12. Procedures Required to Complete this Investigation
13. Experimenting with the Operation of Time and Length
14. The Paired Relation of Time and Length
15. Relative Motion Generates Relative Experiences
16. The Direction of Development
17. The Priority of the Present Tense
18. The Formulas that Define the Three Tenses

It is a mistake in scientific exchange to present only the beautifully polished and objection-free results, as in an art exhibit . . . The creator will all too easily be inclined to show productions complete and without error, and to conceal gaps, uncertainties, and disharmonious contradictions in his scientific knowledge. In so doing, he does an injustice to the feeling for the genuine process of scientific research.

WILHELM REICH (1)

This chapter presents the results of an orgonometric investigation that stretched over eight years. Its aim was to find the "unknown" common functioning principle of the pair of variations, *time* and *length*. Though the aim was clear, there were many unexpected hurdles along the way. I was repeatedly surprised by the complexity of the standard ideas I had to consider. Some of the existing paths I followed simply turned out to be functionally misleading, whereas others were only marginally related to the functions of *time* or *length*. Because these errors still persist elsewhere, it seems appropriate to describe the *partly right* "historic" stages of my investigation, as well as the "polished" results.

The problems I encountered were mainly concerned with the meaning of *time*. As the investigation progressed, I realized our understanding of *time* must be distorted, if it is based upon a misinterpretation of what we perceive — like the old mistake of interpreting the perceived "motion of the sun" as the sun rotating around the earth. Thereafter, when I grasped why objective perceptions can sometimes lead to misinterpretations of natural phenomena, I was set free to make the leap of thought that yielded the answer to the source of *time* and *length*.

The first ten sections of this chapter express views and concepts that are functionally unresolved. Their content should be regarded as *partly right*, even though they contain ideas and observations that opened the way to the resolution of our subject. Sections 11 through 15 constitute the orgonometric core of the text. They present the conclusive orgonometric equations that show the functional source of *time* and *length*, and the paired operation of the two dimensions. In the last three sections, the functional meaning of previously discussed subjects are revised and corrected, according to the new understanding of *time* and *length*.

1. Summary of Previous Observations and Conclusions

Long before I began to investigate this subject with the aid of orgonometry, the process of creating original works of art had led me to observe some of the functional characteristics of *motion*, *time*, and *length*. My earlier observations and conclusions can be summarized as follows:

- In practice, *time* converts into *length*, and *length* converts into *time*, even though the two dimensions are perceived to be different kinds of expression. They are also simultaneously identical, according to some deeper function.
- Everything is perpetually in *motion*, whether or not we perceive it that way. Nothing is really stationary, even in the fixed relation of a tandem motion. Therefore, *motion* must be a basic expression, i.e. a property of the primary function itself.
- When a uniform motion is used to simulate time, it moves in a direction opposite to the accepted "direction" of *time* as represented by the so-called "arrow of time."
- Although a uniform motion can be used to simulate and measure *time*, *time* itself is not a *motion*.[1]

[1] This original thought was forgotten and only recalled in the last stages of my investigation (see section 8).

- We live, act, and perceive in the *present*, therefore, the concept of the *present* is functional and meaningful. But the function and meaning of the concepts *past* or *future* are not clear.

2. First Formulation of the Development of Time and Length

After I became familiar with orgonometry, the special connection between the dimensions of *time* and *length*, previously noted, was now understood to be a functional relation of paired variations (2). But I also recognized their functional status and began viewing them as the most basic of all dimensions. This view was logical, according to my first assumption about the depth of their *common functioning principle* (CFP). I had postulated that *primary motion* is the CFP of the two dimensions. Therefore, both dimensions must also be "primary" quantities. This view is expressed in my first experimental formula for *time* and *length* (June 1982),[2] which can now be represented as follows:

TIME
MOTION ⊰
LENGTH
14.01

This development equation presents the function **MOTION** as the CFP that determines the dimensions of **TIME** and **LENGTH**. Accordingly, the quantitative expressions of **TIME** and **LENGTH** are identical with respect to the function **MOTION**.

Let us accept this temporary arrangement as our first reference equation. If this equation is correct, then *time* and *length* must derive directly from primary *motion*, which estab-

[2] This date, and others like it, refer to the dates these entries were made in the author's workshop notebooks.

lishes their priority as "primary" quantities. Initially, I felt sure this must be the case, since I agreed with the accepted view that *time* flows on inexorably, like primary motion itself.

My earliest doubts about the above equation arose from a thought about the domain of *length*. How can the dimension of length exist in a realm where there are no "fixed" reference points, i.e. the primary realm? And if *length* is not a primary expression, what becomes of the relation of *length* and *time*? Can it be that equation 14.01 is so generalized it expresses a mixture of domains? Months later, after numerous thought experiments, it also became obvious, in the context of this development equation, that even the term "motion" might be too general to represent the correct CFP of *time* and *length*. This one formula opened up so many questions, I could foresee it would lead me on a long orgonometric journey of discovery.

According to the above formulation, the dimensions of *time* and *length* are related as a pair of functional variations. Until we know exactly how they interact, this paired association can be represented in the general form of simple variations as follows:

TIME ⊹ LENGTH 14.02

Let us also accept this general version of the paired variations as our second reference equation.[3]

3. Dimensions Also Express an Order of Function

Eighteen months later, I had a spontaneous thought about the quality of dimensions in general, which turned out to be

[3] A diagrammatic formula and its description, which was published in 1982, independently supported my view of *time* and *length* as a functional pair (3).

important for the further development of my investigation. I suddenly realized *dimensions* must also express a functional order or practical hierarchy. Even though dimensions are manmade distinctions, they all are derive from the *extent* of objective functions, and so they too must express the same functional hierarchy as do their sources — a functional order of dimensions (November 1983). This thought proved to be more penetrating than my earlier assumption about fundamental and non-fundamental dimensions.

In the pendulum experiment, Reich found a way to replace the expression of *mass* with the deeper and simpler expressions of *length* and *time* (4). Thus, if *mass* can be described in terms of *length* and *time*, then the dimension of *mass* must be a narrower functional expression than the dimensions of *length* or of *time*. Most of the familiar dimensions are functionally narrower than *length* or *time*. The dimensions of *area* or *volume* are functionally narrower, and even the dimension of *space*, like the dimension of *volume*, is reducible to the two basic dimensions, *time* and *length*. Indeed, if we can show that *space* is a narrower and more limited functional quantity than either *length* or *time*, then the frequently used *space-time* reference will be seen to be yet another example of an unresolved mixture of functional domains.

A series of orgonometric experiments, i.e. experiments of thought, which followed my new insight about the order of dimensions, was mostly inconclusive. Soon afterward, I tried to construct a table of familiar scientific dimensions, organized according to the order of their derived functional development. This practical exercise directed me correctly, because it strongly highlighted what continued to be *unknown*, namely, the domain (or the domains) of *time* and *length*. Without answers to these basic functional questions, there was no way to complete my table of the "orders of dimensions."

4. Are Time and Length Primary Expressions?

When I attempted to construct a hierarchical table of dimensions based upon the functional hierarchy of their derivations, I was immediately confronted with the following questions about the two most basic dimensions:

- Do the dimensions of *time* and *length* originate in the primary realm?
- Are they both derived from the primary function?
- Is one of them primary (viz. *time*) and the other secondary (viz. *length*)?
- Can it be that both quantities originate in the secondary realm and both derive from some secondary realm function?

The questions were clear, but I had no convincing answer for any of them. Six months later, work on the logic of primary realm functions unexpectedly introduced a way to answer them.

According to the logic of primary functions, all primary functions are *homogeneous* (5). As a pair, *time* and *length* seem to be qualitatively *heterogeneous*, and if so, the pair could not be the expression of a primary realm function. Logically, they would have to be secondary expressions. Once I accepted this possibility, other supportive observations followed rapidly. For example, the measurement of *time* and *length* is associated with *uniform motion*, that is, the motion of straight lines and stable frequencies. (Devices like clocks and tape recorders provide convincing proof of this association.) By contrast, all the variations of primary motion move in irregular curves and spontaneous wavelike paths — quite the opposite of *uniform motion* (September 1984).

The supposition that the two dimensions have a secondary realm origin initiated numerous orgonometric thought experiments. Gradually, I became convinced *time* and *length* are, indeed, derived from some basic secondary function. This conviction was reinforced by concurrent work on the *primordial universe* (chapter twelve). In a mass-free universe, there are no reference points for any form of measurement. In this universe, as in the mass-free primary realm, nothing is structuralized; therefore, there is nothing to anchor our perceptions of *time* or *length*. We can only measure the motion of the primary function through its direct influence upon secondary functions, e.g. its influence upon temperature, the motion of a pendulum, or the pulsation of a living organism. Only matter, i.e. secondary function, provides the kind of "fixed" references needed to measure *time* or *length*. This means that our perceptions of *time* and *length* depend upon the existence of *matter*. Thus, we are led to conclude even *time* itself must have had a beginning, and this beginning occurred after the appearance of the first products of the *function of creation*.

On their own, none of the above observations constitutes proof that *time* and *length* are derived from a secondary function. Together, they become a convincing body of functional evidence, but not yet the proof we seek. Until we have formulated the whole development of *time* and *length*, and discovered the function that determines the association of the pair, we cannot be sure we have resolved anything about our subject.

5. The Error of the "Fourth Dimension"

Time is also known as the "fourth dimension." This is because four measurements are required to completely quantify a "point" or object in "space-time." The other three spatial dimensions are variously described as *length*, *distance*, *width*, *breadth*, *depth*, or *height*. Despite this lexicon of English synonyms, each name represents the same functional quantity, namely, a measure of *length*. However, the so-called "fourth dimension," *time*, is a different kind of quantity. Thus, according to their function, the meaning of the four dimensions of "space-time" can be reduced to only two kinds of function, *length* and *time*.

The practice of referring to *time* as the "fourth dimension" is very misleading, because it does not recognize that dimensions have a functional order, or hierarchy (see section 3). The term "fourth dimension" sustains the illusion that each of the four dimensions is a distinct functional entity and each is of equal importance (December 1984). This mistaken idea has led to the development of a "purely" mathematical technique that extends the number of "space-time" dimensions to ten, or even twenty-six, instead of the usual four — as in string theories (6). Apparently, this multi-dimensional approach has been effective in solving some problems of mathematical physics. But this must mean all ten (or twenty-six) "dimensions" can also be reduced to the same two basic kinds of quantity, *length* and *time*.

Including more numbers (or measurements) in a formula might indeed promote a more accurate quantitative answer. However, this purely technical procedure illustrates how little we can depend upon the *consistency* of abstract systems (e.g. algebra or arithmetic) and their capacity to represent natural functioning. We can generalize from this, where the whole system is *partly right*, as is the case for all systems based in "pure" logic, that each individual expression must initially be regarded as *partly right* (7).

6. Past, Present, and Future

Before I began this orgonometric investigation, I was already convinced by previous experience that only the *present* exists. I considered the *present* to be a concept with a direction generally represented toward the right ⟶, like the direction of a functional *development*. This *direction* coincides with the agreed direction of all functional processes. Later, I came to understand that the functional concept of *development* and, therefore, the *direction* of development, was more basic than my adopted idea of the "direction" of the *present*.

Existing structures contain the record of their past development — from their beginning to the *present*. But their *presence* or *function* does not express any "reverse" characteristic. Similarly, past events that we see now, such as the formation of a star or a galaxy of stars, express themselves now as a present development (⟶). What we observe is the active transmission of radiations of various kinds, which arrive here *now*. Technically, the study of the cosmos is the study of developments occurring *now*. We are not actually looking backward when we observe a *present* consequence of a transmission (June 1985).

However, we are also able to analyze the past, structuralized development of an existing structure from the "frozen" clues within the structure. This "past" is not only the antithesis of development (or the *present*) but is also unlike development (or the *present*). This means that the expression of the "past" and the *present* are functionally *heterogeneous*. Likewise, with transmissions from the "past," the "past" events we *now* observe in the night sky occurred so long ago that everything may have changed at their origins. This "past" is just as obviously an abstraction, since it is analyzed from a presently developing response to a transmission.

The "future" is clearly a non-functional abstraction. "When you get there, there isn't there any more" (due to Gertrude Stein). From the viewpoint of function (or the *present*), the "future" has no real existence. We create the "future" out of our *present* experience and a "memory" of the "past," that is, out of sense perception and structuralized knowledge. According to the logic of functions, we are not looking in the direction ⟶ of development when we look "into the 'future.'" Again, as for the abstraction of the "past," we find ourselves looking backward ⟵.

7. Observations that Contradict the "Arrow of Time"

As soon as I began composing electronic music with tape recorders, I noted that the direction of the tape motion, which simulates the "flow of time," and the direction of the "arrow of time," which is used to represent the "flow of time," contradict each other. After briefly investigating this matter, I simply concluded the direction of the tape movement must be correct, since it fulfills its function. I also noted that the concept of the "arrow of time" is either entirely false or very limited.

In the course of the present investigation, I re-examined my earlier conclusions about the "arrow of time." Again I found, in every practical application, that the direction of the "flow of time" (⟵) appeared to be opposite to the direction (⟶) of the development of functions and opposite to the direction (⟶) of the "arrow of time."

Consider the basic operation of a tape recorder in its playback mode, as illustrated by the following diagram:

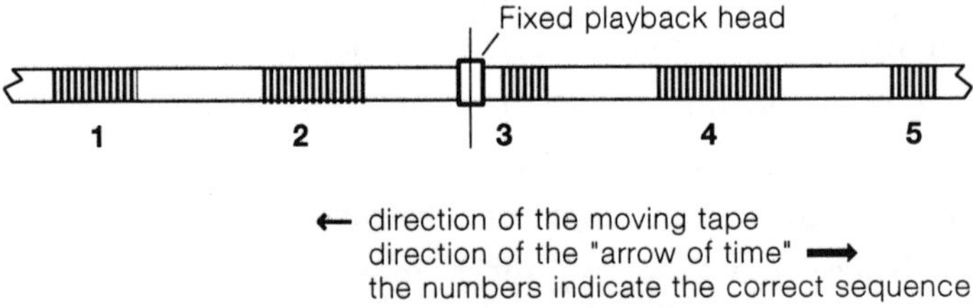

Figure 1. A portion of a prerecorded tape with the magnetic imprints made visible, as it moves past the playback head.

The tape moves *past* the head assembly, which is fixed, in a direction quite opposite to the direction of the "arrow of time," i.e. toward the left ←. Even if we were to change the mechanism so that the head does the moving (→) and the tape remains stationary, the net result would still be the same — the relative movement would make the tape appear to move *past* the head. Functionally, this cannot be otherwise, since the tape can never move *ahead* without altering the original sequence of the recording. Descriptively, the "flow of time" always proceeds from the "future" toward the "past."

With respect to the common motion of the recording and the tape, its direction toward the left ← expresses a past, structuralized development, which is a correct functional description of any kind of record, including printed texts and writing. This directional characteristic is consistent with the orgonometric meaning of the direction toward the left (8).

Consider the same portion of recorded tape dissociated from the tape recorder:

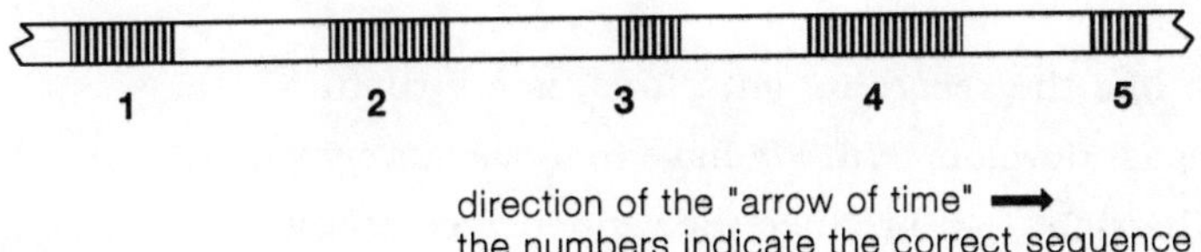

Figure 2. A portion of a prerecorded tape with the magnetic imprints made visible.

Simply by looking at the imprints from left to right →, we can read the imprints in the same *sequence* they were recorded, as indicated by the given numbers. The tape is stationary but our eyes move — equivalent to a moving playback head. If the tape is moved from right to left ← while we fix our gaze on one spot — equivalent to a fixed playback head — we will see the imprints passing by in the same sequence they were recorded.

Both of the above movements produce the relative motion required to reproduce the correct imprint sequence. Therefore, it is still not clear which of the two directions represents the "direction" of *time* or, alternatively, which of the two movements represents the "motion" of *time*.

The same problem of the "arrow of time" also manifests itself in standard time-based graphs where the "arrow of time" points toward the right. Consider, for example, the graph of a series of musical sounds as represented by their sound-envelopes:

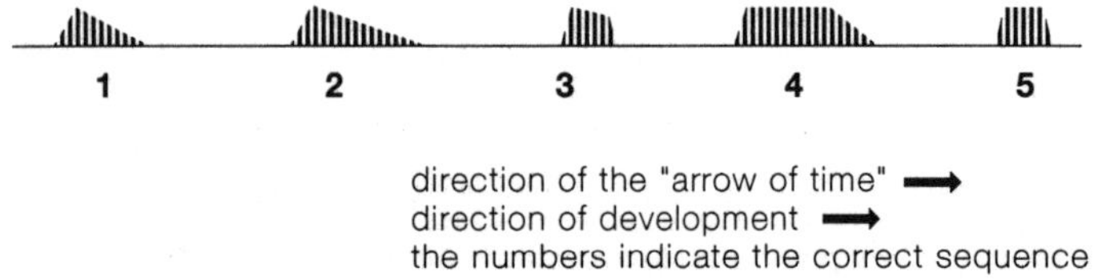

Figure 3. Graph of a sequence of sound envelopes.

This graph, like the recording on a tape, is a structuralized record of a past development. We have to move our eyes from the left to the right to reproduce the sequence correctly. But again, this means that relative to *our* movement, the graphic sequence (i.e. the record) "flows" toward the left ⟵.

Thus, the so-called "arrow of time" does not represent the past, structuralized development of a record. Nor, apparently, does it represent the "direction" of *time*. From the given example (figure 3), we are forced to conclude the "arrow of time" simply indicates the correct way to read the *sequence* of the sound envelopes (September 1987).

8. Time is Not a Motion

In the course of this investigation, I reaffirmed an earlier observation that no matter how we define the meaning of *time*, the functional tense is always the *present*. The universe functions in the *present* (section 6). I also noted, in a preliminary formulation, that *time* and *length* are simple variations of the expression of motion (equation 14.01). The logic of these thoughts, combined with my observations about the error of the "arrow of time," revived a much earlier conclusion, namely, that *time* does not move in any way — it is not another "form of motion" (9). Functionally, *time* is a *static* concept. It seems to be a motion only because it is derived from the experience of *motion*.

On this occasion, I was more prepared to assimilate the thought of *time* as a "fixed" or "non-moving" entity, because it instantly resolved the question of the "direction" of *time* (section 7). *Time* does not "move," and so it does not express a "direction." Following this sudden resolution, I spontaneously recognized that the dimension of *time* is a structuralized concept, just like its paired dimension, the structuralized concept of *length*.

Now I understood why the mechanical function of the tape recorder can be used as functional proof of this observation. The playback head, like *time*, is "fixed" or "stationary," and the tape is propelled past the head, thereby generating the required *relative motion*. The response of the playback head is always the *present*, in relation to the magnetic variations imprinted on the moving tape surface. The playback head "perceives" a constant "flow" (like we perceive the "flow of time") and "responds" to the alternating magnetic changes (like we respond to the alternating changes of physical events).

The error of equating ourselves with machines is still so commonplace that I continue to be very careful about using mechanical examples for any purpose. Nevertheless, machines that function must express their functioning principles correctly, otherwise they would not function at all. Those engineers and inventors who create functioning machines must, with respect to their task, come to think functionally. The time-simulating mechanism of a tape recorder resolves all the basic questions about the relation of *time* and *length* with respect to *motion*. Furthermore, this mechanical solution supports my conclusion that *time* and/or *length* are stationary measures of *motion*.

9. Thoughts about Errors Led to the Answer

One morning when I was busy working on an unrelated task, my thoughts returned to our subject and spontaneously followed a logical path that culminated in the orgonometric formulation of *time* and *length* (October 1987). These thoughts occurred so rapidly, they almost seemed to be simultaneous, and I came to a new conclusion before I could grasp the logic of the sequence. Nevertheless, I will attempt to summarize the specific thoughts that were somehow fused in the creative moment of a thought process:

- *The error of the isolated view of time*: When the expression of *time* is viewed in isolation, it appears to operate like a form of *motion*. But *time* is not a *motion*, even when viewed as an isolated, universal expression. The operation it represents is functionally identical to the qualitative concept of the present — *the state of time is always the present*. This means *time* does not "flow," as I previously thought.
- *The error of giving time (and length) priority over motion*: The classic view of *time* and *length* presumes these two quanti-

ties are more basic than *motion*, and furthermore that each of these quantities can be used to describe the extent of a motion. However, this view is functionally erroneous. The quantity *time* and the quantity *length* are determined by *motion*, and therefore, neither one of them can describe *motion*, whereas *motion* alone can describe *time* and/or *length*.

- *Correcting my working formula*: I reconsidered equation 14.01 and concluded the original CFP labeled "motion" had to be altered to represent some form of secondary motion. If *time* and *length* are secondary realm expressions, then their CFP must also be a secondary function — a form of secondary motion. This also means *time* and *length* only began to exist after the appearance of *matter*, that is, as by-products of the *function of creation*.

10. A More Accurate Development Formula

It follows from the preceding discussion that the function in the role of the CFP, which directly determines the quantities of *time* and *length* (previously labeled as *motion* in equation 14.01), is not a primary realm function but a secondary realm function. We can either describe this secondary function as *regulated motion*, or more simply, as *secondary motion*. With the latter label, we can reconstitute the original development formula (equation 14.01) to reflect the narrower realm of secondary function as follows:

14.03

This development equation shows how the function of **SECONDARY MOTION** determines the paired dimensions **TIME** and **LENGTH**.

The variations of **TIME** and **LENGTH** are identical with respect to the function of **SECONDARY MOTION**. This development equation now supersedes equation 14.01.

However, as I wrote this equation, I also realized that its CFP, "secondary motion," though correct in a general way, is not the specific CFP of *time* and *length*. I could now predict the exact CFP would be one of the known basic processes or operations of secondary motion. I knew I was close to the correct answer. Then, for no apparent reason, I began to think about the special theory of relativity, Einstein's third paper (1905), in which, as I recalled, he demonstrated that *time* is a *variable* quantity — not "absolute" or invariable, as was previously assumed. He had postulated the velocity of light is always the same, and all other quantitative expressions are *relative*. My thoughts seemed to leap a gap. Eureka! That is the functional answer! "Relative motion" is the CFP of *time* and *length*!

11. The Source of Time and Length — the Whole Formula

We can now present the source of *time* and *length* in the form of a qualitatively whole development equation:

14.04

The equation states that **RELATIVE MOTION** determines the pair of dimensional variations **TIME** and **LENGTH**. The dimensions of **TIME** and **LENGTH** are identical with respect to the function of **RELATIVE MOTION**. We also see that the extent of a **RELATIVE MOTION** can be defined by two artificial distinctions, the quantities of **TIME** and **LENGTH**.

According to this equation, *relative motion* is the common functioning source for our perception of *time* and our perception of *length*. The equation demonstrates that the quantity of *time* and the quantity of *length* coexist. Though we readily distinguish these two quantities, what they each express should not be viewed as independent of the other, i.e. as isolated quantities. Since both are derived from one unified functional expression (*relative motion*), they can be correctly understood only as simultaneous expressions or properties. Together, *time* and *length* constitute the *whole* quantitative expression of *relative motion*. Therefore, to describe the *whole* quantity of *relative motion*, both measures must be recorded.

12. Procedures Required to Complete this Investigation

The resolution of development equation 14.04, the source of *time* and *length*, does not conclude our orgonometric investigation of this subject. There are still some important questions to be answered and a lot of loose ends left over from the contents of the earlier stages of thought, described in sections 2 through 7. There is also a need to confirm, or functionally prove, that the new equation — a product of thought creation — is indeed qualitatively correct.

Since we know what we need to uncover and consider, the next phase of our investigation can begin. Based upon the given form and content of equation 14.04, we can plan the order of further research. Such an orderly approach may help some researchers, but in my experience, no procedural order is better than any other. Nevertheless, it is useful at this point to outline what we do *not* know, that is, what still needs to be resolved.

Technically, the first orgonometric requirement would be to explore the relation of the paired variations, *time* and *length*. Knowing how the paired quantities operate would complete the functional information represented in equation 14.04. As a second step, we would examine the meaning of the CFP, *relative motion*, and reach a preliminary understanding of the functional origin of this form of motion. Thereafter, we would consider the general and detailed implications of the new finding. This step could well lead us toward a deeper finding or a further, more extensive, development formula. Progress of this kind would then constitute one form of orgonometric "proof." Last, we would apply our new comprehension to some actual example and show how the new finding has altered our interpretation of the function of the example. This too constitutes another form of orgonometric "proof."

My investigation did *not* proceed according to the technical order described above. After years of intermittent work on this subject, I suddenly and rapidly seemed to resolve the crucial formula (equation 14.04) — too suddenly for me to accept immediately. Consequently, I first needed to consider how the meaning of the new formula would alter my earlier observations and thoughts. To do this I had to assume, for a while, that the new formula was accurate. Later I turned to examine the operation of the paired variations and the other considerations described above. Since there is no advantage in repeating my procedure, I will now continue according to the technical order outlined above.

13. Experimenting with the Operation of Time and Length

At the start of this chapter, we presented the paired relation of *time* and *length* in the general orgonometric form of simple paired variations (equation 14.02). This formula continued to be *right,* until the discovery of the CFP defined the two dimensions as secondary realm expressions. Before this was clear, there was no need (and no way) to reconsider this reference equation. While *time* and *length* appeared to be "primary realm" expressions, they could only be related as simple variations, since all primary expressions are simple variations of each other (9).[4] But now, given that *time* and *length* are secondary realm expressions, we need to investigate how the two quantities relate to each other.

Time and *length* are abstract concepts, and both are measured according to agreed manmade units and subunits. As "pure" concepts, their paired relation is indeed that of *simple variations*, since all "pure" concepts are simple variations of each other.[5] In this specific context, equation 14.02 continues to be meaningful and correct. However, the concepts of *time* and *length* are not "pure" abstractions, because they derive from common perceptions of a concrete function, namely, the function of *relative motion.* Both dimensions represent real, objective properties, and we are concerned here with this functional meaning or content.

The orgonometric technique for discovering the operation of a given pair of functions is to test the given pair in all possible paired settings (10). To begin, let us assume the two expressions, *time* and *length,* are functionally *homogeneous*, that is, of the same kind. Their relation to each other could therefore be one of the following: *Simple variations*, *antagonistic opposites*, *alternating opposites*, or *mutually attractive opposites.* None of these paired operations seems to describe the relation

---

[4] Perhaps I might add, I never reconciled my perceptions of *time* and *length* as heterogeneous qualities with the technical thought that they could be *simple variations.*

[5] To be presented at a later date.

of *time* and *length* in the context of the development of *relative motion* (equation 14.04), or according to the way we perceive them.

Next, we assume the two expressions are functionally *heterogeneous*, that is, of a different kind. This is exactly how they are perceived — quite possibly the strongest kind of confirmation. Also, the different etymology of the names supports this view, though we cannot be too sure about this source of support. Then, there is the operation of functional *transformation* between heterogeneous functions of the same domain. We have already noted that *time* can be described as a *length* and vice versa. Given the operation of a two-way transformation, together with the other supporting observations described above, we conclude that *time* and *length* constitute a *heterogeneous* pair of variations (July 1989).

14. The Paired Relation of Time and Length

The basic dimensions *time* and *length* express different kinds of quantity. They are, therefore, qualitatively *heterogeneous*. Since each quantity can be technically transformed into the other, their paired relation operates as a two-way transformation. Two equations are required to represent this operation as follows:

$$\text{TIME} \nrightarrow \text{LENGTH} \qquad 14.05a$$

$$\text{LENGTH} \nrightarrow \text{TIME} \qquad 14.05b$$

These equations show that the quantity **TIME** can be transformed into a quantity of **LENGTH**, and the quantity **LENGTH** can be transformed into a quantity of **TIME**.

Practical applications have demonstrated the validity of the above equations. For hundreds of years, *time* has been measured with a scale of *length* (e.g. the circular scale of a clockface), and astronomical *length* has been measured with a scale of *time* (e.g. the scale of light-years). However, the transformations in the above equations represent an operation limited to the manmade system of measurement, which means it describes a technical transformation. Such transformations do not represent the subjective way we perceive the extent of *relative motion*. **TIME** remains *time*, even when it is measured as a distance, and **LENGTH** remains *length*, even when it is measured as a duration.

We should also note the above equations illustrate a subtle, practical difference between a function and our sensory response to the function. According to the logic of functions, *relative motion* expresses its *quality* as a whole. Therefore, its *extent* must also be whole. Yet, in this case, we perceive its *extent* as two different kinds of quantity, *time* and *length*. The split in the expression of *relative motion* can be traced to the "response of the living." Life has developed a variety of senses, and this sensory differentiation leads us to perceive such differences. Whenever we study the *properties* of a single function, we need to remind ourselves repeatedly about the unity of these properties (e.g. the unity of *time* and *length*), if only to avoid the serious functional error of viewing any functional *property* as an isolated entity.[6]

15. Relative Motion Generates Relative Experiences

Relative motion is a natural, ubiquitous expression of all secondary realm functions within the moving streams of the

[6] *Properties* are artificial distinctions of a function and because of this they are always individually incomplete or *partly right*. This observation will be more fully discussed in chapter fifteen.

primary energy "ocean." According to equation 14.04, the quantities of *length* and *time* are the perceived extent of this motion. However, there are many situations where *relative motion* is not experienced as a *motion* but as a "frozen" or *structural* relationship, such as a fixed *location*, a defined *space*, or the *size* of an object. Unfortunately, the everyday experiences of such structured relations appear to support an isolated view of *length* — where *length* has nothing to do with *time*. Conversely, there are other daily situations where *relative motion* is experienced as a *uniform motion*, and this consistent quality, which can be used to measure time, is mistaken for a "characteristic" of *time* itself, that is, as the "motion" of *time*. This idea supports an isolated view of *time* — where *time* has nothing to do with *length*. We can now see how the experience of *relative motion* and the non-functional way we interpret this experience have repeatedly misdirected our thoughts about the meaning of *time* and the meaning of *length*.

In practice, both kinds of measurement, *time* and *length*, are required to describe the *whole* quantity of *relative motion*. This functional requirement continues, even in such cases where the *relative motion* is not perceived or detected, or where it appears to be "absolutely" uniform.

The *relative motion* of matter and secondary energy is influenced by the motion of differentiated primary energy streams. This does not mean that primary motion directly determines *relative motion*. We know the *function of creation* intervenes, and according to the logic of functions, this must somehow be reflected in the development of *relative motion*. However, until its functional origin is fully understood, we shall continue to describe *relative motion* as a secondary realm function. For the present, this will distinguish it from the operation of differentiated *primary motion*.

Relative motion and the dimensions of *time* and *length* are concepts that have no meaning in the context of the *primordial universe.* In this realm of function, there was nothing that can be described as "fixed" or "frozen" — no structures and no structuralization (11). However, the *function of creation* altered the primordial condition by introducing structural functions in the form of *matter* and *secondary energy.* Even the appearance of a few such functions presents the "fixed" or "frozen" qualities that anchor our perceptions of *time* and/or *length.* Therefore, I conclude that the beginning of the quantities *length* and *time* follows the earliest creations of secondary matter and its consequence, *the beginning of relative motion.*

16. The Direction of Development

We can now answer the questions raised about the representation of *time* by means of an arrow, as in the example of a graph of a musical sound sequence (figure 3 in section 7). The contradiction described, with respect to the accepted direction of the "arrow of time," is indeed significant as equation 14.04 confirms. The quantity *time,* like the quantity *length,* is a structuralized abstraction. Since *time* does not move, the "movement" mistakenly ascribed to *time* is actually one of the two different ways we perceive the extent of *relative motion.* As a structuralized concept, the only "direction" associated with *time* is the abstract direction toward the left ⟵, which is opposite to the direction of the "arrow of time." Thus, the "arrow of time" completely misrepresents the functional meaning of *time.*

The *direction* of development is an inherent characteristic of every functional development. Development proceeds in only one direction, that is, toward the differentiation of functions and toward the greater complexity of narrower functions and their various functional interactions. Thus, the orgonometric meaning of the "direction of development" is essentially a qualitative description of the one-way transformation that occurs in the development of differentiated functions. In orgonometry, we agree to represent this *directional quality* with an arrow ⟶ pointing toward the right (12).

The invention of the *direction of development*, which occurred more than five years after the initiation of orgonometry, is one of Wilhelm Reich's major thought creations. It replaces the confusing idea of the "arrow of time" (now shown to be erroneous) with the definable, qualitative *direction* of the development of functions. Having formulated equation 14.04, we now understand the orgonometric *direction of development* represents a quality functionally deeper or more fundamental than *time*. The resolution of our subject confirms the primacy of the *direction of development*, because without it, I would not have been so certain about the contradiction between the direction of the tape movement and the so-called "direction of time." I can now foresee how this one invention could help us dissolve some of the time-related contradictions that have emerged in modern theoretical physics, such as the mathematical "reversibility of time," or the "irreversibility of entropy."

The concept of the "arrow of time" is correct only insofar as it represents the *direction of development*, a *quality* that has nothing to do with the *quantity* of time. We should simply discard the whole concept of the "arrow of time" and replace it with the orgonometric *direction of development* ⟶.

17. The Priority of the Present Tense

The concept of the *tenses*, which is so clearly articulated in language, expresses the logic of "pure" thought. According to the logic of functions, of the three tenses, only the concept of the *present* is relevant. The concepts *past* and *future* have little if any functional meaning, since they are abstractions derived from abstract generalizations, i.e. "pure" abstractions. Neither one exists except within the limit of the operations of thought — a limit frequently encountered in manmade systems, such as language, mathematics, science, and so on. Together, the three tenses constitute a functional contradiction or mixture of unrelated functional domains.

The quality of the *present* is functionally deeper than both the qualities of the *future* and the *past*. Again, we can use the operation of a tape recorder to illustrate the functional difference between the *present* and the other two abstractions. The following diagrammatic sequence shows the *relative motion* of a length of tape as it approaches (A), contacts (B), and passes (C), the playback head of a tape recorder:

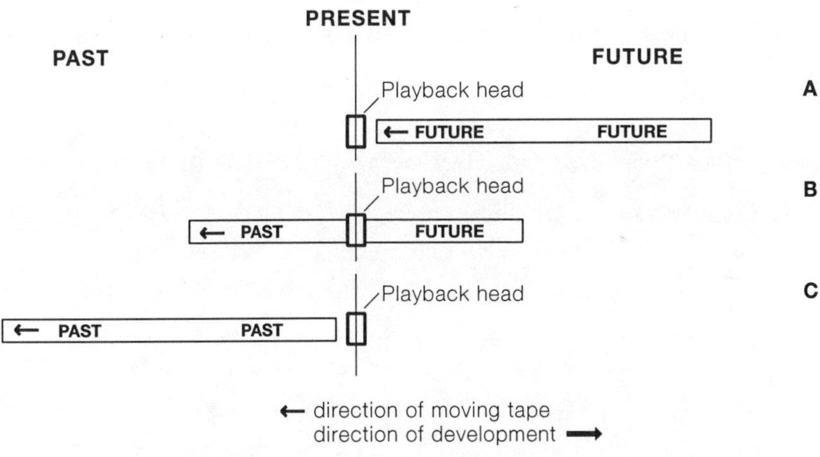

Figure 4. Three phases of a moving portion of tape in relation to the playback head of a tape recorder.

This illustration shows the operating similarity of the concepts of the **FUTURE** and the **PAST**. Both are directed toward the left

←; both can be represented by the same structure or substrate (the tape); and a predetermined **FUTURE** (i.e. an existing tape recording) abruptly changes from **FUTURE** to **PAST** as it moves *past* the playback head. The **PAST**, represented by an existing tape recording, also changes into the **FUTURE** when the playback operation is initiated. This functional alternation, or "flip-flop," is a clear indication that **PAST** and **FUTURE** are indeed *homogeneous* abstract concepts.

The illustration also shows the very different character of the concept of the **PRESENT**. The expression of the **PRESENT** is represented by the function of the playback head — not by the tape, the movement of the tape, or the direction of the tape. The playback head, with its associated electronic processors, functions like a mechanical equivalent of a sense organ. Within its limit, this mechanical sensor responds to the *relative motion* of the tape and simulates the *living* perception of the **PRESENT**. The playback head defines the **PRESENT** as a product of the *sensation of relative motion.*

18. The Formulas that Define the Three Tenses

Having concluded that the concept of the *present* tense represents an objective human experience of a concrete function, my curiosity directed me to find the paired variation of this experience. Given the previously demonstrated fact that the *present* is derived from a sensation of *relative motion*, the required variation must also be derived from a sensation generated by *relative motion.* The answer spontaneously emerged a few days later, namely, the objective human experience of *existence.* These constituents can now be arranged in the form of a development equation as follows:

14.06

The equation shows that the sensation of **RELATIVE MOTION** generates two kinds of related human experiences, the **PRESENT** and **EXISTENCE**. The experience of the **PRESENT** and the experience of **EXISTENCE** are identical with respect to the sensation of **RELATIVE MOTION**.

The *present* and *existence* represent different kinds of qualities. In practice, we find the quality of the *present* can be described in terms of *existence*, and *existence* can be described in terms of the *present*. These technical transformations support the conclusion that the two qualities constitute a *heterogeneous* functional pair. This paired relationship can be reformed as follows:

the PRESENT ⇸ **EXISTENCE** 14.07a

EXISTENCE ⇸ **the PRESENT** 14.07b

The operations of equations 14.07a and 14.07b are similar to the paired equations for *time* and *length* (see section 14). In both cases, the individual constituents of the pair represent a "part" of the whole function. Such distinctions are inherently artificial, even though they have a significant functional content, and even though we do experience the determining function (or CFP) in this divided way.

In contrast to the above, the concepts of the *past* and the *future* are individually derived from other concepts. Accordingly, they cannot be directly associated with a concrete function and must be defined as creations of thought. Their conceptual origins can now be presented in the following two orgonometric arrangements:

the PRESENT
DEVELOPMENT ≻⇸ **FUTURE** 14.08

The equation shows how the concept of the **FUTURE** is derived from the perceived association of the functional concepts of the **PRESENT** and of **DEVELOPMENT**.

14.09

The equation shows how the concept of the **PAST** is derived from the perceived association of the functional concepts of the **PRESENT** and of **STRUCTURE**.

According to equation 14.09, the concept of the *past* cannot be derived in the *primordial universe* because of the absence of material *structure*. The same is not true for the concept of the *future* (equation 14.08), which can be derived in the *primordial universe*. In yet another way, these orgonometric observations confirm: The concept of the *past* only applies to secondary realm functions; and the *past* has a distinct beginning, like the quantity of *time*.

15

ADVANCES IN THE BASIC TECHNIQUE

Sections

Do not try to hide your mistakes, speak about them frankly, and be proud of knowing your mistakes. Do not try to be perfect. Your mistakes are your most reliable signposts on your road.

WILHELM REICH (1)

All the developments of thought presented in this book can be ascribed, in large part, to the influence of the technique of orgonometry. Before I encountered this technique, my capacity for original, functional thinking was sporadic and restricted. There was always "something" about the practice of functional thinking that seemed to elude me. Reich's paper on orgonometry expanded my grasp of functional thinking extensively, and I immediately began to use the technique in a creative way. That experience together with subsequent experiences, many of which are reflected in the content of this volume, convinced me of the benefits of using the technique of orgonometry.

Presenting this technique of thought has become, for me, a profoundly important act. This book began with a complete exposition of the technique of orgonometry, and it will conclude with an updated discussion of my work-related contributions to the original technique.

The contents of the first ten sections of this chapter should be regarded as supplements to related sections in "Basic Orgonometry" (chapter two). They present the correct solutions to some subtle, technical inconsistencies that appeared in the original text. The subsequent sections include discussions of techniques and technical details not discussed in "Basic Orgonometry," because they were deemed to be obvious at the time. Since then, experience has shown these procedures need to be described more fully. In the concluding sections, we review two practical, orgonometric tools that evolved out of my work, and we end with a note about the functional view of equations and equality.

1. The Direction of Development[1]

Our investigation of the true nature of *time* reveals that the concept of "the flow of time" is erroneous. Further, the *direction* of the "arrow of time" has nothing to do with time but is, in fact, a misconceived description of the *direction of development* (2). Therefore, these *partly right* concepts should not be used in the orgonometric definition of the *direction of development.* The present version of "Basic Orgonometry" (chapter two, which is otherwise reproduced without significant changes) eliminates all previous references to the "flow of time."

To say *time* is incorporated in the meaning of *development* limits the whole meaning of development. Though it is true that *time* is associated with the development of secondary functions, it should not be included in the meaning of primordial or primary development. *Time* did not exist in the *primordial universe*, nor is it expressed in *primary development*. Thus, the concept of *time* is excluded from the deepest (or broadest) meaning of the *development* of functions.

2. Primary Development[2]

Reich's equation for *primary development* (equation 2.11) is a crucial generalization that ultimately led him to the formulation of the *function of creation* (equation 2.13). Unfortunately, the original *primary development* formula introduced some technical inconsistencies that also became incorporated in the

[1] Supplement to section 2.11.

[2] Supplement to section 2.16.

original equation for the *function of creation*. One of these inconsistencies became apparent when I was investigating the *primordial universe*, and because of this, I introduced an "improved" version of the original equation in my presentation of this investigation (equation 12.09).

Soon after I completed that chapter, I recognized a further inconsistency in both the original and in my "improved" version of the general equation for *primary development*. A functional conclusion in my research notes made this plain. All primary variations are *homogeneous* qualities. *Heterogeneity* only appears after the development of secondary functions, such as *matter* and *secondary energy*. Therefore, in accordance with the orgonometric technique of letter-labeling, even the lower-case reference letters "x" and "y" are inconsistent in the context of primary function formulas.

In abstract functional equations, *homogeneous* functions are represented by the same letter-label. Since each of these functions is also different, we simply distinguish them with a reference number. For example, two *homogeneous* functions labeled **A**, become **A2** and **A3**. The letter-label is selected to express the quality of the function it represents, whereas the number-label is qualitatively "neutral" — merely used for reference.

To construct a very generalized equation, we would still follow the same procedure. However, we also need to generalize the number-labels that, by convention, brings us back to the use of the letter "n." Accordingly, we can represent the *primary development* of functions with the following generalized equation:

V1 ⊀ **Vn1** / **Vn2** 15.01

where **V** represents the principle of variation, **n** represents a generalized reference number, and **1**, **2**, the specific constituent references.

The equation states that the *primordial variation*, **V1**, develops the paired variations, **Vn1** and **Vn2**. Functions **Vn1** and **Vn2** are therefore identical with respect to their CFP, **V1**.

V1 is the abstract representation of the primordial variation, the CFP of all nature — previously labeled **N. Vn1** and **Vn2** represent the first two primary variations and all other pairs of primary variations, which developed thereafter — previously labeled **Vx** and **Vy**.

3. The Function of Creation[3]

Since the general expression of *primary development* is an integral part of the *function of creation*, the technical changes in the above equation must also appear in an updated formulation of the *function of creation* as follows:

$$\mathbf{V1} \; \not\prec \begin{matrix} \mathbf{Vn1} \\ \mathbf{Vn2} \end{matrix} \not\succ \; \mathbf{A1} \tag{15.02}$$

where **V1** represents the primordial variation, the CFP of all nature — previously labeled **N. Vn1** and **Vn2** represent the first two primary variations and all other pairs of primary variations, which develop thereafter — previously labeled **Vx** and **Vy**. **A1** represents the new, unique functional product of creation, which emerges from the fusion of two specific primary variations, **Vn1** and **Vn2**.

Because all primary constituent functions are letter-labeled **V**, the *homogeneous* quality of the development of primary functions, which precedes the operation of fusion, is immediately apparent in this version of the *function of creation*. We can

[3] Supplement to section 2.19.

readily see how the *function of creation* "transforms" primary functions into a secondary function distinguished by its letter-label **A**. The new, unique product of creation (a secondary realm proto-function) may itself become the CFP of a further secondary development. As the origin of a secondary development, this proto-function is rightfully labeled **A1**, i.e. a prime secondary function.

NOTE: The way Reich originally derived the equation for the *function of creation* is a matter of historic record. This was briefly described in section 2.19. Thus, we should continue to show the original forms of equations 2.11, 2.12 and 2.13, along with the updated versions.

4. Orgonometry and Orgonomic Functionalism[4]

All products of creation are identical with respect to the mass-free operations of the *function of creation*. Within this primary context, the creation of *orgonomic functionalism* and the creation of *orgonometry* are indeed simple paired variations.

However, when creation occurs within a secondary environment, the new product is subject to a process of natural regulation, whereby it eventually becomes integrated with the environment. The way the new product is altered by this interactive process establishes the potential of the whole end product. In the case of the creation of *orgonometry*, the new technique emerged from the regulating process as deeper, sharper, and more integrated than the method of *orgonomic functionalism*. Orgonometry is very well aligned with the processes it describes. It is not confined by the operations of verbal language or by the limit of "pure" thought.

[4] Supplement to section 2.21.

5. The Two Directions of Research[5]

Research played a major role in the development of the technique of orgonometry. Much of this research was original, that is, directed toward the CFP of a functional development. This abstract direction is the "reverse" of the functional *direction of development.* When working in this contradictory direction, we are led into making "reverse" observations that are functionally erroneous and practically confusing. We need to be alert to this source of error when the direction of our work contradicts the *direction of development.*

Since the technique is now given, we have less reason to present it in any way other than aligned with the *direction of development.* With this consistent approach, we can initially avoid the kind of confusion the "reverse" direction introduces. In our presentation of "Basic Orgonometry," nothing forced us into thinking about functions in a backward direction, except the need to cover this subject. Reich, the discoverer, was often confronted with the problem of "reverse" observations and, as a consequence, wrote about it extensively.

6. Reconstruction of a Past Development[6]

The development of *armor*, equation 2.18, is derived from a formulation by Reich, but it differs from the original in a way that should have been noted in "Basic Orgonometry." The original was not constructed with orgonometric symbols. Normally, this would mean the arrangement is a diagram and not an orgonometric equation.

Did Reich intend this to be a diagram, or can we say the missing graphic components of the functional symbols were simply a printer's error?

---

[5] Supplement to sections 2.25, 2.26, and 2.27.

[6] Supplement to section 2.28.

In the original text, the *armor* "diagram" is followed by an abstract representation of the same development, intended to show the type of whole integration involved. This abstract version is a completely regular orgonometric equation, which convinced me the missing components of the functional symbols were indeed a printer's error. Consequently, I reproduced Reich's formula as an orgonometric equation.

A few years after "Basic Orgonometry" was published, I needed to review the whole of Reich's original discussion and formulation for a quite different reason. It concerned the use of the term "armor" and whether this was a valid label for a specific function. During this review, I once again observed the diagrammatic appearance of the original formula. On this occasion, the discrepancy startled me, because I had grown accustomed to using my orgonometric version as if it were the established representation of the *function of armor*. My response reminded me I had inadvertently "finalized" Reich's original formula by omitting to note this discrepancy in the text of "Basic Orgonometry."

7. Opposite Concepts[7]

With due respect to Reich, section 2.29 has generated the most questions and contradictions. Many experiments with *concepts* in orgonometric settings led me to invent specific ways to formulate the functional content of related *concepts*. Coincidentally, this activity expanded my grasp of the functional content of concepts in general. The entire subject will be presented at a later date, but my technical advances can now be applied to the *concepts* discussed in section 2.29 of "Basic Orgonometry."

[7] Supplement to section 2.29.

The *primordial function* is the deepest source for the qualities of *finity* and *infinity*. At this level of function, our view of these two qualities becomes greatly simplified. They appear to be nothing more than *properties* of the *primordial function*, which makes them somewhat artificial — "parts" of a whole function. As *properties*, each is technically unitary, or *singular*, because each represents the whole function that is its source. This conclusion is supported by the fact that the two qualities, *finity* and *infinity*, coexist even though they are extreme opposites. Furthermore, primordial properties (or qualities) pre-exist the process of *development*, therefore each primordial property must be a simple variation of the *primordial function* itself.

Many useful concepts are functionally *partly right* (3). Differentiating the *properties* is a good example of a useful, yet *partly right*, operation of thought. The concepts *infinity*, *determinism*, and *time*, for instance, are derived from concrete functions, but each is also an artificial abstraction of a *whole* function — an isolated view. To some extent, conceptual distinctions are a consequence of the natural development of differentiated, sensory organs. But the major problem here is the loss, or diminution, of our feeling of "being at one with the universe." To understand the whole, we must either *feel* integrated or deliberately remind ourselves of how such distinctions are integrated. Therefore, technical formulations of *properties of a function* should always include, in some way, the expression of the integrating function.

The singular quality of *finity* and the singular quality of *infinity*, which reflects the singularity of the *primordial function*, can be more meaningfully represented in the following precisely defined monolinear form:

$$\mathbf{V1} \,\textbf{⨍}\, 1 \,\textbf{⨍}\, \infty \qquad 15.03$$

where **V1** represents the *primordial variation* — the *whole* function, **1** represents *finity*, and ∞ represents infinity — the *isolated* properties. Compare this arrangement with the original equation 2.19.

The same functional observation applies to the opposite concepts of *determinism* and *freedom*. Each of these concepts can also be traced back to a property of the *primordial function*. Again, each is a singular functional expression, like the *primordial function* itself. These two opposite concepts can also be better represented in the following precisely defined monolinear form:

V1 ⫞ DETERMINISM ⫞ FREEDOM 15.04

where **V1** represents the *primordial variation* — the *whole* function, and **DETERMINISM** and **FREEDOM** are the *isolated* properties. Compare this arrangement with the original equation 2.20.

The form of the above equations (15.03 and 15.04) unifies the artificially differentiated constituents and their functional source (**V1**). It also conveys the simultaneity of all the constituent qualities, which is a specific characteristic of *properties*.

Without digressing to describe all my abortive experiments with the original form of the *history equation* 2.21, I eventually concluded that the concept "history" is too artificial to function as the CFP of the given equation. Further experiments with other, similarly artificial concepts led me to construct a general formula for all such concepts, equation 4.01 (4). Consequently, we can reconstruct the *history equation* correctly as follows:

15.05

The paired relation of **secondary DEVELOPMENT** and **STRUCTURE** is perceived and then synthetically integrated with the generalized concept of **HISTORY**.

We have noted elsewhere, according to the orders of function, that the expression of *development* is deeper and broader than the expression of *structure* (5). *Development* is a primary expression, whereas *structure* is a secondary expression. Although they may be paired, *development* and *structure* constitute a mixed domain pair (6). We can correct this hierarchical mixture by limiting the meaning of "development" with a qualification, namely, *secondary development*. In his text, Reich also distinguishes two levels of development: Development in general, which originates with the *primordial function*, and development that can become *structuralized*.

The logic of the above arrangement is further supported by the following observations about the operations of *concepts*. According to the above equation, the concept *history* is functionally very indirect, since it is derived from two derived concepts, i.e. it is two stages removed from any objective function. *Secondary development* and *structure* are also concepts, but they are *directly* derived from observed functional processes. Thus, on the basis of objective content (i.e. the "reality principle"), the concepts of *secondary development* and *structure* take precedence over the concept of *history*.

8. Quality and Quantity as Functional Properties[8]

Some time after resolving the *concept of history* formula above, I noticed the arrangement of the formula for *functional properties* (equation 2.22) expresses a similar error of thought.

[8] Supplement to section 2.34.

The concept "functional properties" is too artificial to function as the CFP of the given objective development. The original equation can be correctly reconstructed as follows:

15.06

As this equation shows, the paired relationship of **secondary QUALITY** and **QUANTITY** is perceived and then synthetically integrated with the generalized concept of **FUNCTIONAL PROPERTIES.**

The appreciation of *quantity* is limited to the realm of *secondary function*, whereas *quality* is a primary expression. Thus, in relation to the derived and artificial concept of *functional properties*, *quality* and *quantity* constitute a mixed domain pair. In the above equation, redefining the pair as *quantity* and *secondary quality* eliminates the earlier mixture of functional domains.

We have demonstrated that *quantity* does not have a meaning within the context of the *primordial universe* (7). The only way we can measure a primary *quality* is *indirectly* through its "effects" upon secondary functions. By contrast, even the deepest *qualities* of primary realm functions can be *directly* perceived. That life responds to *qualities* directly and first is clearly no accident but an expression of the operation of the primary function within the plasma of living organisms. We need to accept the priority of our bio-sensory system as a tool of scientific research before we can hope to penetrate or recognize the deepest levels of function. Thus, the practical priority orgonometry assigns to *qualitative observations* and *qualitative formulations* is further confirmed by the logic of this observation (8).

9. The Function that Pairs with Entropy[9]

In the middle of his discussion of opposite concepts, Reich introduced a very significant scientific equation for comprehending the principle of *entropy* and its relation to the function of primary energy. I omitted this equation in my section on "opposite concepts," because it seemed to be an unnecessary digression within the given context. However, since I intended to reproduce all the equations that appear in Reich's paper on orgonometry, I feel I should correct this one omission here.

Given that every function is paired with another function (except the primordial function), even the function of *entropy* or "running down" must be paired with a function of "building up." We can expect to find, somewhere in nature, a function that *builds up a potential.* Unless we can prove that *entropy* is a singular expression of the primordial function — which it is not — the logic of functions leads us to look for the functional pair of *entropy.* Reich's discovery of orgone energy completed our picture of the development of this pair of functions as follows:

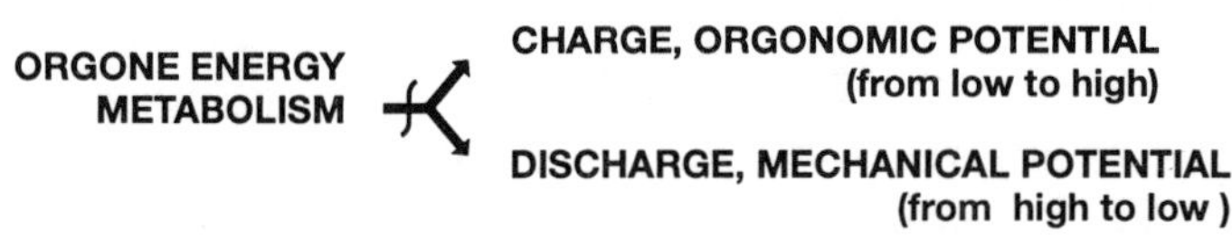

15.07

The equation states that the **METABOLISM** of **ORGONE ENERGY** determines both the **CHARGE** and the **DISCHARGE** of distinctly different orders of potential — **ORGONOMIC POTENTIAL** and **MECHANICAL POTENTIAL** respectively.

---

[9] Supplement to section 2.29.

10. Three Forms of an Orgonometric Equation[10]

The three formulations shown below are generalized with respect to their constituent functions. Nevertheless, the constituent labels are fundamentally correct as abstract descriptions of primary functions.

The form of paired functions:

$$\mathbf{Vn1} \nrightarrow \mathbf{Vn2} \qquad 15.08$$

where **V** represents the principle of variation, **n** represents a generalized reference number, and the numbers **1** or **2**, the practical constituent reference. Thus, **Vn** represents all primary variations other than the *primordial variation*.

The form of the development of functions:

$$\mathbf{V1} \nrightarrow \begin{matrix} \nearrow \mathbf{Vn1} \\ \searrow \mathbf{Vn2} \end{matrix} \qquad 15.09$$

where **V1** represents the singular *primordial variation*.

The form of the function of creation:

$$\mathbf{V1} \nrightarrow \begin{matrix} \nearrow \mathbf{Vn1} \\ \searrow \mathbf{Vn2} \end{matrix} \rightarrowtail\!\nrightarrow \mathbf{A1} \qquad 15.10$$

where **A1** represents the product of creation, the new CFP that emerges from the *fusion* of **Vn1** and **Vn2**.

[10] Supplement to section 2.35.

11. The Technique of Abstraction

The practical advantage of introducing the technique of orgonometry with abstract formulations seems obvious. It allows us to illustrate the common patterns of the operations of functions without being sidetracked by the individual characteristics of any one function. Further, it excludes any confusion of concepts that name-labels might introduce.

However, when we investigate the interactions of specific functions and their development, we usually think of each quality or function by name. As a consequence, the majority of our experimental formulations and the majority of the formulations presented in this book are word-labeled orgonometric equations.

Letter-labels are more general, yet they convey something basic and immediate about the kinds of constituent functions word-meanings are now too complex to convey.[11] As previously noted, letter-labels distinguish between functions of the same kind and functions of a different kind — between *homogeneous* and *heterogeneous* pairs of variations (9). We need to make this distinction before we can abstract a word-labeled equation correctly. More often than not, this simple distinction is easily made, although here our perceptions can sometimes mislead us. For example, *time* and *length* are perceived to be a *heterogeneous* pair, whereas qualitatively, they must be regarded as *simple variations* because they are the properties of a single function, *relative motion* (10).

Any given word-labeled equation can be abstracted and reviewed in this more basic form. There is always a possibility we can learn something new about a process by simply converting the word-labeled formula into a letter-labeled for-

[11] When the language was new, words and their meanings were well-integrated, according to the characteristics of the functions they represented.

mula. The abstract version of the *orgasm formula*, for example, immediately revealed why the sequence of functions in the cycle of the *orgasm formula* is so specific (11).

12. The Technique of Monolinear Translations[12]

Any given development can be formally rearranged as a monolinear statement. For example, a basic two-stage development equation:

$$\mathbf{A1} \nprec \begin{matrix} \mathbf{A2} \\ \mathbf{A3} \end{matrix} \qquad 15.11a$$

can be rearranged as follows:

$$\mathbf{A1} \nrightarrow (\mathbf{A2} \not- \mathbf{A3}) \qquad 15.11b$$

A three-stage development equation:

$$\mathbf{A1} \nprec \begin{matrix} \mathbf{A2} \nprec \begin{matrix} \mathbf{A4} \\ \mathbf{A5} \end{matrix} \\ \mathbf{A3} \nprec \begin{matrix} \mathbf{A6} \\ \mathbf{A7} \end{matrix} \end{matrix} \qquad 15.12a$$

can be rearranged as follows:

$$\mathbf{A1} \nprec \begin{matrix} \mathbf{A2} \nrightarrow (\mathbf{A4} \not- \mathbf{A5}) \\ \mathbf{A3} \nrightarrow (\mathbf{A6} \not- \mathbf{A7}) \end{matrix} \qquad 15.12b$$

or as follows:

$$\mathbf{A1} \nrightarrow \{[\mathbf{A2} \nrightarrow (\mathbf{A4} \not- \mathbf{A5})] \not- [\mathbf{A3} \nrightarrow (\mathbf{A6} \not- \mathbf{A7})]\} \qquad 15.12c$$

---

[12] First introduced in sections 6.2 and 6.3.

Equations 15.11b and 15.12c are the monolinear representations of the two given development equations. Equation 15.12b is a transitional arrangement. The braces, brackets, and parentheses have been introduced to retain, as far as possible, the functional hierarchy expressed in each development equation. They define the boundaries of each domain in relation to its CFP or realm. This hierarchical information would be lost without the use of brackets, and even with the brackets, the order of the domains is not immediately apparent in a monolinear representation. For example, in equation 15.12c, **A3** is far removed from **A2** (its functional pair). In this form of notation, **A3** appears *after* **A5**, which belongs to the narrower third domain of function. We can see how easily the original order of the development can be confused by the omission of a single bracket or by a minor disarrangement in the sequence of the functions.

Word-labeled formulas can also be reconstructed as monolinear equations. In this form, a word-labeled formula resembles the organization of a sentence (12). It will demonstrate how best to describe the given formula in words. Reconstructing a functional process as a monolinear equation is another useful tool for re-examining the operation of the given process.

13. Letter-labels

As noted in the preface to "Basic Orgonometry," I carefully reproduced Reich's equations exactly as he formulated them. In some instances, the constituent functions are represented with lower-case letter-labels, whereas in others (the majority), they are represented with upper-case letter-labels. Since I could not find a significant reason for this notational practice,

I decided to continue using capital letter-labels exclusively, until a reason emerged to justify introducing lower-case letters.

New observations about the properties of functions, as well as completion of my work on *length*, *time*, and *relative motion*, uncovered a significant reason for introducing lower-case letter-labels.

The properties of any given function are subordinate to that function. The function does not split into a series of *properties* just because we can abstract these qualities or quantities. The *properties* have no independent existence. Each property represents the whole function, yet none, on its own, can *express* the whole function. A function expresses all of its properties simultaneously and together, and nothing less than *all the properties* can express that function fully. Therefore, we cannot isolate a property from its function without generating a serious error of thought.

In abstract formulations, we can differentiate between *whole functions* and the *properties* of a function by consistently using upper-case labels for the former and lower-case labels for the latter. We will reserve the use of lower-case letter-labels for representing *quantities* (or *qualities*) that have a subordinate meaning. For example, *time* is subordinate to the function *relative motion*; *infinity* is subordinate to the function *primordial variation*.

After I started using lower-case letter-labels to represent the *properties* of a function, I soon found this notational expression provides other, unforeseen, practical benefits. For instance, we can combine upper-case and lower-case labels to represent an abstracted property in an undistorted, non-isolating, compound label, which includes a reference to the whole function.

For example, the development of *relative motion* (equation 14.04) can be abstractly represented as follows:

$$\mathbf{R} \prec \begin{matrix} \mathbf{Rt} \\ \mathbf{Rl} \end{matrix}$$ 15.13

where **R** represents *relative motion*, **Rt** represents *relative time*, and **Rl** represents *relative length*. The compound label **Rt** conveys the inseparable bond between the dimension of *time* and its source. Similarly, **Rl** conveys the bond between the dimension of *length* and its source.

With the aid of compound letter-labels, the deepest, or broadest, meaning of the concepts of *finity* (**1**) and *infinity* (**∞**) can be abstractly represented as follows:

$$\mathbf{V1} \prec \begin{matrix} \mathbf{V1} \\ \mathbf{V\infty} \end{matrix}$$ 15.14

where **V1** on the left represents the *primordial variation* — the source for the deepest meanings of the paired concepts *finity* and *infinity*. **V1** on the right is a compound label representing *primordial finity*. It conveys the inseparable relation between the *primordial function* and the deepest meaning of the concept of *finity*. Similarly, the compound label **V∞**, *primordial infinity*, intrinsically represents the deepest concept of *infinity*.

Compare the above formulation with the original equation 2.19 and my updated version 15.03. Note how the system of compound letter-labeling introduces, in this one case, another way to represent the *primordial function* **V1** in the abstract, namely, **V∞**.

Compound labeling practice is also consistent with the way we modified the constituent labels of the generalized equation for *primary development* (15.01) and the equation for the *function of creation* (15.02).

14. Names that Represent Functions

The name of a process seldom represents the function that determines the development of the process. For instance, the name "development" does not represent the specific function that determines a *development of functions*; "history" does not determine *development* or *structure*; and "evolution" does not determine *variations in the environment* or *variations in the living response*. In recent times, researchers have distinguished, described, and named numerous natural processes without knowing their functional source (i.e. the CFP). We cannot use this kind of name as a functional label, because such names are limited by how they were derived. The names of many diseases exemplify what we mean. They are particularly useless as function-labels, because they convey nothing about the operation of the disease they represent.

When working with orgonometric formulations, check every name-label to be sure it represents a definite function — a quality that *works*, or a quality that *can determine another function*. We can easily be misled by an inappropriate name masquerading as a function. In my experience, the technique will eventually help to expose this kind of error, but not without a persistent effort of inquiry.

Occasionally, we may need to use more than one word to represent a given function (equations 13.14 and 15.07). When this becomes necessary, try to be concise. Wordy name-labels tend to obscure the functional meaning of the whole equation. The aim should be to make the formulation sharp, immediate, and easy to read.

15. The Development Template[13]

The introduction of a distinctive symbol to represent any *unknown* function prepared the way for creating a very useful orgonometric tool, the *development template*. This device describes the operations of a functional development without expressing the *quality* of any constituent function. We specify its extent by stating the number of domains. A three-domain *development template* appears as follows:

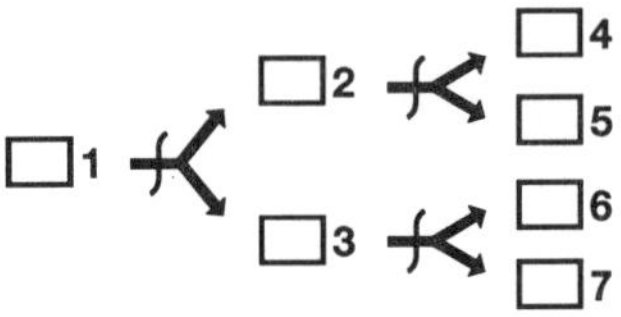

□4
□2
□5
□1
□6
□3
□7 15.15

| 1 | 2 | 3 | domains of function

This arrangement reproduces the pattern of a three-stage development equation. Since all the constituent functions are represented by the symbol for an unknown function □, the whole arrangement consists of nothing more than symbols. In this state, the device is not an equation but rather a pattern or *template* for an orgonometric equation.

This orgonometric tool or template can now be filled in or adjusted to suit a variety of different tasks. Obviously, the task itself must be generally understood before a specific template can be prepared. For examples of its uses, see chapters ten and eleven.

The *development template* allows us to represent the correct order of known and *unknown* functions within a development without involving the *direction* of research. This is specially advantageous where the *direction of research* contradicts the

[13] This technical tool was first introduced in chapter eleven.

direction of the process being investigated. Also, it is particularly useful for describing or investigating the correct order of *mixed domain* formulations (13).

16. The Present State of Primary Development[14]

The present state of primary development is indefinably extensive. The generalized form of the equation for primary development given previously, equation 15.01, does not represent the present state in a sufficiently detailed way. To express this extensive process succinctly, yet meaningfully, we have devised a special variation of the development symbol. The new symbol, in conjunction with an abstract technique for indicating a very large reference number, provides us with the following practical formula:

$$\mathbf{Vn^{10}{+}1} \;\prec\; \begin{matrix} \mathbf{Vn^{10}{+}2} \\ \mathbf{Vn^{10}{+}3} \end{matrix} \quad \begin{matrix} \mathbf{Vn^{20}{+}4} \\ \mathbf{Vn^{20}{+}5} \\ \mathbf{Vn^{20}{+}6} \\ \mathbf{Vn^{20}{+}7} \end{matrix}$$ 15.16

where **V** represents the principle of variation, **n^{10}+1** represents the whole constituent reference, a very large, undefined number, and **+1**, **+2**, etc., are the practical reference numbers for each constituent.

The equation represents an extensive primary development as it appears now, many stages after the beginning of development. Function **Vn^{10}+1** dissociates into the paired functions **Vn^{10}+2** and **Vn^{10}+3**. These in turn dissociate through many stages, producing the numerous variations that presently exist. These are represented on the right of the formula by a

[14] This equation for primary development was introduced in chapter twelve.

sample of only four constituents with extremely large reference numbers, **$Vn^{20}+4$**, **$Vn^{20}+5$**, **$Vn^{20}+6$**, and **$Vn^{20}+7$**.

Primary variation **$Vn^{10}+1$**, the CFP of the above equation, is many stages of development removed from the *primordial variation* **V1**. Nevertheless, the two functions are related as simple variations. Every one of the primary constituents shown is a simple variation of the *primordial variation.* Each expresses all the properties of the *primordial variation.* They are differentiated mainly by their scale and secondary associations, including their location and indirect secondary expressions.

The usefulness of the above formula will become self-evident when we eventually present an account of our investigations of the *function of creation.*

17. Equations and Equality

Although we call our orgonometric formulations and arrangements "equations," we generally reserve the term "equality" for describing a quantitative identity. The meaning of "equality" has become so closely identified with *quantity* over the past two hundred years that the word has presently lost its original, *qualitative* meaning.

Equality, in a functional sense, is more deeply rooted than the *function of creation.* Its roots are to be found in the *primordial universe* itself. All primary variations, from the cosmic to the plasmatic (indifferent to measures of scale or extent), are all identical yet simultaneously different. To convey this deeper meaning of "equality," we generally use the term "identical" or "functionally identical" and often follow this with a reference to the CFP that determines the given association.

Every "equal" relationship is determined by the CFP of the variations being considered. If we search deeply enough for their CFP, all living things can be shown to be functionally identical — plants and animals. With the ultimate function, the *primordial variation* (**V1**), in the role of the CFP, all functions become ultimately identical yet simultaneously different.

In research and for practical applications, we are much more concerned with comprehending the interactions of closely related constituent functions. Here, we search for the immediate functional pair and the immediate function that determines the operation of the pair. When these are found and confirmed, the whole equation defines the function of the given CFP. If the given CFP can be represented by a *quantity*, then the whole equation can be reconstructed as a *quantitative functional equation*. Here, we can introduce some of the standard procedures and terminology of algebra, including the terms "equality" and "equals," without verbally misrepresenting the relation of the functional *quantities*.

NOTES

References to the text are identified by a number that combines the chapter number with the section number. For example, section 8.04 *means: Chapter 8, section 4. Equation numbers follow the same system. Thus,* equation 13.18 *can be found in chapter 13. Cross references within the text are identified in the same way.*

Preface

1. Reich, W.: *People in Trouble*. Rangeley, Maine: Orgone Institute Press, xiv, 1953.

Chapter 1: Enter the Logic of Functions

1. Reich, W.: "Rules to Follow in Basic Research," *Orgone Energy Bulletin*, 3(1):64, 1951.

Chapter 2: Basic Orgonometry

1. Sharaf, M.: "Further Remarks of Reich: Summer and Autumn, 1948," *Journal of Orgonomy*, 5(1):102, 1971.
2. Reich, W.: "Orgonomic Functionalism," *Orgone Energy Bulletin*, 2(1):2-6, 1950.
3. Reich, W.: "Complete Orgonometric Equations," *Orgone Energy Bulletin*, 3(2):65-71, 1951.
4. Reich, W.: "The Orgone Energy Charged Vacuum Tubes," *Orgone Energy Bulletin*, 3:235-266, 1951.
5. Reich, W.: "Orgonometric Equations: I. General Form," *Orgone Energy Bulletin*, 2(4):182, 1950.
6. Ibid., 183.
7. Reich, W.: *Cosmic Superimposition*. Rangeley, Maine: Orgone Institute Press, 17-26, 44-63, 1951.
8. Reich, W.: *The Cancer Biopathy*. New York: Orgone Institute Press, 51-63, 1948.
9. Grad, B.: "Wilhelm Reich's Experiment XX," CORE, 7:130-145, 1955.

10. Reich, W.: "Experimental Investigation of the Electrical Function of Sexuality and Anxiety," (1937), *Journal of Orgonomy*, 3(2):132-154, 1969.

Chapter 3: The Real Meaning of E = mc²

1. Reich, W.: "Orgonomic Functionalism," *Orgone Energy Bulletin*, 2(1):3-4, 1950.
2. Reich, W.: "Complete Orgonometric Equations," *Orgone Energy Bulletin*, 3(2):70-71, 1951.
3. Reich, W.: "Orgonomic Functionalism," *Orgone Energy Bulletin*, 2(1):3-4, 1950.
4. Review sections 2.08 to 2.10.
5. Review section 2.12.
6. Review sections 2.23, 2.24.
7. Review section 2.10.
8. Reich, W.: "Orgonometric Equations: I. General Form," *Orgone Energy Bulletin*, 2(4):167, 1950.
9. Dietz, D.: "Atomic Bomb," *Encyclopaedia Britannica*. Chicago: University of Chicago Press, 2:647B, 1946.

Chapter 4: Function Viewed as a CFP

1. Reich, W.: "Orgonometric Equations: I. General Form," *Orgone Energy Bulletin*, 2(4):170, 1950.
2. Review sections 2.14, 2.15, 3.22.
3. Review sections 2.25, 2.26, 2.31.
4. Review equations 2.07, 2.08, 2.10, 2.11.
5. Review sections 2.23, 2.24.
6. Reich, W.: "Orgonomic Functionalism," *Orgone Energy Bulletin*, 4:193-194, 1952.

Chapter 5: The Formulation of *Right* and *Partly Right*

1. Reich, W.: *The Function of the Orgasm*. New York: Noonday Press, 7, 1961.
2. Review sections 2.23, 2.24.

3. Reich, W.: "Orgonomic Functionalism," *Orgone Energy Bulletin*, 2(2):49-52, 1950.
4. Review sections 2.06 to 2.10.
5. Review section 4.01.
6. Review section 2.16.
7. Reich, W.: *The Function of the Orgasm.* New York: Noonday Press, 48, 1961.
8. Placzek, B. R. edit: *Record of a Friendship.* New York: Farrar, Straus & Giroux, 418, 1981.
9. Reich, W.: *Ether, God, and Devil.* Rangeley, Maine: Orgone Institute Press, 48, 1949.
10. Reich, W.: "Orgonomic Functionalism," *Orgone Energy Bulletin*, 2(3):106, 1950.
11. Review section 2.32.

Chapter 6: Language and the Development of Functions

1. Reich, W.: *Ether, God, and Devil.* Rangeley, Maine: Orgone Institute Press, 9, 1949.
2. Review sections 2.12, 2.13.
3. Reich, W.: "Orgonomic Functionalism," *Orgone Energy Bulletin*, 2(1):11, 1950.
4. Reich, W.: "Complete Orgonometric Equations," *Orgone Energy Bulletin*, 3(2):66, 1951.
5. Reich, W.: *Contact With Space.* New York: Core Pilot Press, 258, 1957.
6. Review section 2.16.

Chapter 7: Art as a Process

1. Klee, P.: *Paul Klee on Modern Art.* London: Faber & Faber, 15, 1957.
2. Review sections 2.27, 2.28.
3. Baker, E. F.: *Man in the Trap.* New York: Macmillan Publishers, 1967.
4. Review section 6.02.
5. Review sections 2.17 to 2.20.

6. Klee, F.: *Paul Klee Gedichte.* Zurich: Verlag Die Arche, 7, 1960.
7. Review section 4.07.
8. Reich, W.: "Orgonometric Equations: I. General Form," *Orgone Energy Bulletin,* 2(4):178, 1950.
9. Kandinski, W.: *Point, Line and Plane.* New York: Dover Publications, 1971.
10. Klee, P.: *Paul Klee on Modern Art.* London: Faber & Faber, 1957.
11. Spiller, J.: *Paul Klee: The Thinking Eye.* London: Lund Humphries, 1961. Includes an alternative translation of the text of Paul Klee on Modern Art.
12. Reich, W.: *The Cancer Biopathy.* New York: Orgone Institute Press, 51-63, 1948.

Recommended Reading:

Gimpel, J.: *The Cult of Art.* London: Weidenfeld & Nicholson, 1969. See chapter 14 for a sharp and exciting description of the initial impact of photography upon painting, and for the individual responses of the great artists of the time.

Chapter 8: Hegel's Dialectic Concept

1. Reich, W.: *Ether, God, and Devil.* Rangeley, Maine: Orgone Institute Press, 78, 1949.
2. Review sections 2.06 to 2.10.
3. Reich, W.: "Orgonomic Functionalism," *Orgone Energy Bulletin,* 2(2):49-52, 1950.
4. Review sections 5.06 to 5.08.
5. Review equations 4.01, 4.02.
6. Review section 4.01 to 4.03.
7. Review section 5.11.

Chapter 9: The Function that Defines the Goal

1. Reich, W.: "Orgonometric Equations: I. General Form," *Orgone Energy Bulletin,* 2(4):178, 1950.

2. Lampkin, J.: "Readers Forum," *Journal of Orgonomy*, 21(2):263, 1987.
3. Review section 5.11.
4. Review equation 2.18 for a more extensive formula.
5. Reich, W.: "Orgonometric Equations: I. General Form," *Orgone Energy Bulletin*, 2(4):168, 1950.
6. Review section 2.26.

Chapter 10: How to Integrate an Unknown Function

1. Reich, W.: "Rules to Follow in Basic Research," *Orgone Energy Bulletin*, 3(1):63-64, 1951.
2. Review sections 2.04, 2.13, 2.14.
3. Review section 5.11.

Chapter 11: The Orders of Function

1. Reich, W.: "Orgonomic Functionalism," *Orgone Energy Bulletin*, 2(3):113, 1950.
2. Review section 2.13.
3. Review section 2.29.
4. Reich, W.: "Orgonometric Equations: I. General Form," *Orgone Energy Bulletin*, 2(4):174, 1950.
5. Review section 10.01.
6. Review section 2.19.

Chapter 12: The Primordial Universe

1. Reich, W.: *Ether, God and Devil.* Rangeley, Maine: Orgone Institute Press, 19, 1949.
2. Reich, W.: "Orgonometric Equations: I. General Form," *Orgone Energy Bulletin*, 2(4):176-177, 1950.
3. Review section 2.16.
4. Review figure 6.04.
5. Review section 11.02.
6. Reich, W.: "The Oranur Experiment," *Orgone Energy Bulletin*, 3(4):267-325, 1951.
7. Reich, W.: "Orgonometric Equations: I. General Form," *Orgone Energy Bulletin*, 2(4):182, 1950.

8. Review section 2.19.
9. Review equation 2.13.
10. Review section 11.02.

Chapter 13: The Function of the Orgasm

1. Reich, W.: "Orgonometric Equations: I. General Form," *Orgone Energy Bulletin*, 2(4):169, 1950.
2. Reich, W.: *The Function of the Orgasm*. New York: Noonday Press, 1961.
3. Reich, W.: "Orgonomic Functionalism," *Orgone Energy Bulletin*, 2(3):99-123, 1950.
4. Ibid. 107.
5. Reich, W.: *The Function of the Orgasm*. New York: Noonday Press, 7, 1961.
6. Reich, W.: "Orgonomic Functionalism," *Orgone Energy Bulletin*, 2(3):105-107, 1950.
7. Ibid. 108.
8. Ibid. 104.
9. Ibid. 116-121.
10. Reich, W.: "Experimental Investigation of the Electrical Function of Sexuality and Anxiety," (1937), *Journal of Orgonomy*, 3(2):132-154, 1969.
11. Reich, W.: "Orgonomic Functionalism,," *Orgone Energy Bulletin*, 2(3):107-108, 1950.
12. Ibid. 108.
13. Review section 2.08.
14. Review sections 5.07 to 5.09.
15. Review section 6.02.
16. Review section 2.09.
17. Review section 2.08.

Chapter 14: The Source of Time and Length

1. Reich, W.: *Ether, God, and Devil*. Rangeley, Maine: Orgone Institute Press, 5, 1949.
2. Review sections 2.06, 2.07.

3. Baker, C.F.: "The Orgone Energy Continuum: The Ether and Relativity," *Journal of Orgonomy*, 16(1):55-58, 1982.
4. Reich, W.: *Contact with Space*. New York: Core Pilot Press, 102, 1957.
5. Review section 12.08, 12.04, 12.06.
6. Hawking, S. W.: *A Brief History of Time*. New York: Bantam Books, 162-165, 1988.
7. Review section 5.11.
8. Review section 2.26.
9. Meyerowitz, J.: "Modulating the Time Analogy," *Analog Sounds*, 5:46-47, 1975.
10. Review sections 2.08, 2.10.
11. Review section 12.05.
12. Review section 2.11.

Chapter 15: Advances in the Basic Technique

1. Reich, W.: "Rules to Follow in Basic Research," *Orgone Energy Bulletin*, 3(1):63-64, 1951.
2. Review section 14.16.
3. Review section 5.17.
4. Review section 4.03.
5. Review section 12.05.
6. Review section 11.03.
7. Review sections 12.05, 14.04.
8. Review section 2.33.
9. Review section 2.07.
10. Review sections 14.14, 15.13.
11. Review section 13.11.
12. Review section 6.04.
13. Review sections 11.03, 11.06.

BIBLIOGRAPHY

Development of Functional Thought

Reich, W.: "Dialectical Materialism and Psychoanalysis," (1929), *Sex-Pol.* New York: Vintage Books, 1972.

___ "Dialectic Materialism as Applied to Research and Theoretical Formulations," (1938), *Journal of Orgonomy*, 13(1):5-30, 1979.

___ "The Dialectical-Materialistic Method of Thinking and Investigation," *The Bion Experiments* (1938). New York: Farrar, Straus & Giroux, 1979.

Orgonomic Functionalism

Reich, W.: *Ether, God and Devil.* Rangeley, Maine: Orgone Institute Press, 1951.
Retranslated and republished, New York: Farrar, Straus & Giroux, 1973.

___ "Orgonomic Functionalism," *Orgone Energy Bulletin.* Rangeley, Maine, 2:1-15, 49-62, 99-123, 1950; 4:1-12, 186-196, 1952.
Retranslated and republished, *Orgonomic Functionalism.* Rangeley, Maine, 1:1-29, 2:1-23, 1990; 3:1-19, 1991; 4:1-18, 1992; continuing.

___ *Cosmic Superimposition.* Rangeley, Maine: Orgone Institute Press, 1951.
Retranslated and republished, New York: Farrar, Straus & Giroux, 1973.

Orgonometry

Reich, W.: "Orgonometric Equations: I. General Form," *Orgone Energy Bulletin*, Rangeley, Maine, 2:161-183, 1950.

___ "Complete Orgonometric Equations," *Orgone Energy Bulletin*, Rangeley, Maine, 3(2):65-71, 1951.

___ *Contact With Space*. New York: Core Pilot Press, 1957. (For examples of applied orgonometry, see pp. 73-78, 95-100, 101-110.)

Meyerowitz, J.: *Basic Orgonometry*. New York: RRP Publishers, 1985.

___ *Applied Orgonometry*. Easton, Pennsylvania: RRP Publishers, issue 1, March 1986; issue 2, November 1986.

___ *Before the Beginning of Time*. Easton, Pennsylvania: RRP Publishers, 1994.

TABLE OF ORGONOMETRIC SYMBOLS

This table includes all the orgonometric symbols used in this book. Reich presented them in papers on Orgonometry, in *Contact With Space* (see Bibliography), and in his introduction to "Melanor, Orite, Brownite and Orene," (McCullough) *Core*, 7:29-31, 1955. Symbols devised by the author are duly noted.

Paired Functions

⨍ General symbol for the operation of paired functions that are identical with respect to a common functioning principle (CFP). Also used as the specific symbol for paired functions that operate as *simple variations*. The latter usage should not be assumed without proof.

→⨍← The operation of *mutually attractive opposite* paired functions.

←⨍→ The operation of *antagonistic opposite* paired functions.

⇄⨍ The operation of *alternating opposite* paired functions.

⨍→ The operation of a *functional transformation* from one function into another. The symbol essentially describes transformations between heterogeneous paired functions of the same domain (i.e. paired functions). However, in linear translations of a functional equation, this symbol can also represent the *development* of a transformation between a CFP and a heterogeneous variation.

The operation of a functional identity between paired functions in a quantitatively formulated equation. It indicates that the *whole* arrangement is both *functionally and numerically* integrated (complete), or in the process of becoming integrated (suspended).

Development of Functions

The basic *operation of development.* Paired functions on the right develop from the CFP on the left.

The operation of an *extensively differentiated development*, where the constituent functions are too numerous to count, e.g. the present state of *primary development* (due to the author).

The operation of a functional development in a quantitatively formulated equation. It indicates that the *whole* arrangement is both *functionally and numerically* integrated (complete), or in the process of becoming integrated (suspended).

The operation of *fusion.* This development follows superimposition in the *function of creation.*

The operation of an *investigative process* directed toward the CFP, a development unique to the operation of thought. This symbol is like the above symbol for *fusion*, but each is distinguished by its context.

Symbolic Notations

ƒ	Abbreviation for the terms *function* or *functional.*
⟶	Represents *development*, and indicates the agreed *direction of development.*
⟵	Represents *structuralized, past development,* and indicates an abstract direction that only exists in the operations of thought.
∫	Sign of a *proto-material* state, as in ∫H (proto-hydrogen), a creation product of atmospheric orgone fusion.
□	Represents an *unknown* function in the context of an orgonometric equation. This symbol is not intended to convey any functional *quality* or *characteristic* (due to the author).
□3	Represents an *unknown* function with a reference number. The number is merely used as a notational device to distinguish each *unknown* constituent in a relationship of *unknown* constituents. In this context, the number conveys no other meaning (due to the author).

INDEX OF INTEGRATED FUNCTIONS

The reference numbers identify the equations that integrate the given function. The number to the left of the point represents the chapter, and the number to the right, its sequence within the chapter. For example, 6.07 *means: Chapter six, equation seven.*

INDEX